*Carbon Adsorption
for Pollution Control*

PRENTICE HALL SERIES IN PROCESS AND POLLUTION CONTROL EQUIPMENT

by Nicholas P. Cheremisinoff and Paul N. Cheremisinoff

Carbon Adsorption
for Pollution Control =

Nicholas P. Cheremisinoff
Paul N. Cheremisinoff

P T R Prentice Hall
Englewood Cliffs, New Jersey 07632

Library of Congress Cataloging-in-Publication Data

Cheremisinoff, Nicholas P.
    Carbon adsorption for pollution control / Nicholas P.
Cheremisinoff, Paul N. Cheremisinoff.
        p.   cm.
    Includes bibliographical references and index.
    ISBN 0-13-393331-8
    1. Pollution. 2. Carbon, Activated.   I. Cheremisinoff, Paul N.
II. Title.
TD191.5.C48 1993
628.5--dc20

                                                    92-23500
                                                    CIP

Editorial/production supervision: *Brendan M. Stewart*
Prepress buyer: *Mary Elizabeth McCartney*
Manufacturing buyer: *Susan Brunke*
Acquisitions editor: *Michael Hays*

The publisher offers discounts on this book when ordered in bulk quantities. For more information, write: Special Sales/Professional Marketing, Prentice Hall, Professional Technical Reference Division, Englewood Cliffs, NJ 07632.

Printed in the United States of America
10 9 8 7 6 5 4 3 2 1

ISBN 0-13-393331-8

Prentice-Hall International (UK) Limited, *London*
Prentice-Hall of Australia Pty. Limited, *Sydney*
Prentice-Hall Canada Inc., *Toronto*
Prentice-Hall Hispanoamericana, S.A., *Mexico*
Prentice-Hall of India Private Limited, *New Delhi*
Prentice-Hall of Japan, Inc., *Tokyo*
Simon & Schuster Asia Pte. Ltd., *Singapore*
Editora Prentice-Hall do Brasil, Ltda., *Rio de Janeiro*

"Concern for man himself and his fate must always form the chief interest of all technical endeavor. Never forget this in the midst of your diagrams and equations."

—Albert Einstein

Contents

**4   *Gas-Phase Adsorption and Air
Pollution Control*   79**

Appendix   Physical Constants and Conversion Factors

Index

Preface

There has been a historical and continuing interest in the use of activated carbon. Carbon use predates Christian civilization with applications including decolorization and purification of water, various products, solutions, and gases. Interest in carbon use for air and water pollution control as well as traditional industrial applications has continued and has received increased attention since the early 1970s with the advent of more stringent environmental regulations.

Adsorption processes are based on the physical properties of specifically prepared solids known as adsorbents which attract selectively and retain chemical compounds on their surfaces. The most widely used adsorbent is activated carbon. Other adsorbents which are significantly less important are activated alumina, silica gel, and molecular sieves. Adsorption equipment functions to bring the process stream and solid adsorbent into direct contact to facilitate adsorption for removal of fluid impurities. In all of these operations the adsorbent is worked on a cyclic basis. Regeneration, stripping of adsorbents, and recovery may be accomplished in a number of ways. Use of these methods and their applications are discussed.

As the previous books in this series, this volume is intended as a practical book that can be used by the practitioner and should also be useful as a basic reference for those who may want to look at carbon adsorption in the field of pollution control. Every attempt has been made to keep the material presented readily understandable to a broad cross section of engineers, scientists, technicians, and managers whose fundamental backgrounds may be widely diverse. Discussions which might be too technical or theoretical in this field have been avoided intentionally and this volume is presented to provide a working knowledge and description of the adsorption process and in particular activated carbon.

—*Paul N. Cheremisinoff*

Carbon Adsorption
for Pollution Control

1

Introduction to Adsorption

Adsorption is the selective collection and concentration onto solid surfaces of particular types of molecules contained in a liquid or a gas. By this unit operation gases or liquids of mixed systems, even at extremely small concentrations, can be selectively captured and removed from gaseous or liquid streams using a wide variety of specific materials known as *adsorbents*. The material which is adsorbed onto the adsorbent is called the *adsorbate*. The two mechanisms involved, chemical adsorption and physical adsorption, focus specifically on carbon adsorption.

When gaseous or liquid molecules reach the surface of an adsorbent and remain without any chemical reaction, the phenomenon is called *physical adsorption* or *physisorption*. The mechanism of physisorption may be intermolecular electrostatic or van der Waals forces, or may depend on the physical configuration of the adsorbent such as the pore structure of activated carbon. Physical adsorbents typically have large surface areas. The properties of the material being adsorbed (molecular size, boiling point, molecular weight, and polarity) and the properties of the surface of the adsorbent (polarity, pore size, and spacing) together serve to determine the quality of adsorption. There are also the following parameters which can be used to improve physical adsorption:

- Increase the adsorbate concentration.
- Increase the adsorbate area.

- Select the best absorbent for the specific gas system.
- Remove contaminants before adsorption.
- Reduce the adsorption temperature.
- Increase the adsorption contact time.
- Frequently replace or regenerate the adsorbent.

Physical adsorption units may either be regenerable types or use disposable adsorbents. Regeneration of physical adsorbents is accomplished via any combination of three mechanisms, namely temperature, pressure, and concentration swings. Freshly regenerated adsorbents theoretically remove 100 percent of the contaminants and at the other extreme significant quantities of contaminants begin to escape at the breakthrough point. Physical adsorption systems may either consist of two beds (adsorption, desorption) or three beds (adsorption, desorption, cooling).

When gaseous or liquid molecules adhere to the surface of the adsorbent by means of a chemical reaction and the formation of chemical bonds, the phenomenon is called *chemical adsorption* or *chemisorption*. Heat releases of 10–100 kcal/g-mol are typical for chemisorption, which are much higher than the heat release for physisorption. With chemical adsorption, which is far less common than physiosorption, regeneration is often either difficult or impossible. Chemisorption usually occurs only at temperatures greater than 200°C when the activation energy is available to make or break chemical bonds.

Activated carbon is the most widely used adsorbent today. It is usually categorized as a physical adsorbent and also as a nonpolar adsorbent. It can be produced from a wide variety of carbonaceous materials and provides an extremely high internal surface area within its intricate network of pores. A total surface area range of 450–1,800 m²/gram has been estimated. Only a portion of that area is available for adsorption in pores of the proper size. For organic solvent adsorption, carbonaceous material is capable of removing at least 85 percent of the emissions. Activated carbon comes in three general types: granular or natural grains, pellets, and powders. The natural grains which are hard and dense are most appropriate for gaseous-phase adsorption applications while other types, liquid-phase adsorbents, are commonly used to decolorize or purify liquids and solutions. Generally, liquid-phase carbons have about the same surface areas as gas-adsorbing carbons, but have larger total pore volumes. Liquid-phase carbons are generally either powdered or granular, the former mixed and later filtered from the liquid, the latter charged into a bed. The variety of activated carbon affects what is adsorbed and how well. Activated carbon with a concentration of small pores tends to adsorb smaller molecules than large-pored carbons. The chemistry of the carbon surface and its ash constituents also affect behavior.

The major application division of the carbon adsorption unit operation is between liquid-phase adsorption and gaseous-phase adsorption. Gaseous-phase carbon adsorption is primarily used for solvent vapor recovery and selective gas separations. Liquid-phase carbon adsorption is used to decolorize or purify liquid, solutions, and liquefiable materials such as waxes and in water and wastewater treatment as a polishing removal or tertiary treatment.

LIQUID-PHASE ADSORPTION

Carbon adsorption from the liquid phase is generally classified as a nonpolar or hydrophobic type of adsorption operation. It is generally used to remove less polar contaminants from polar bulk streams. The two basic liquid-phase equipment designs for carbon adsorption are the fixed-bed and pulsed-bed arrangements. The equipment makes use of either powdered or granular liquid-phase carbons. Fixed-bed equipment can assume the form of either single or multiple columns which can operate in series, in parallel, or both.

Contact-Batch Operation

In the typical batch configuration for a contact-batch operation, the equipment consists of an agitated tank constructed of materials suitable for the liquids being processed. The agitation allows the carbon particles to continually contact fresh portions of liquid causing mild turbulence. Where materials are sensitive to oxidation which can be caused by excessive mixing, the adsorption should be conducted under a partial vacuum or an inert atmosphere. The liquid-carbon mixture is pumped from the tank through a filter (commonly a plate and frame type). Powdered carbon is normally applied as a slurry to minimize dusting problems.

Fixed Single-Column System

A single-column system for liquid-phase carbon adsorption is used in situations where the following conditions prevail. Laboratory testing has indicated that the breakthrough curve will be steep; the extended lifetime of the carbon at normal operating conditions results in minor replacement or regeneration costs; the capital cost of a second or third column cannot be justified due to insufficient savings in carbon cost; to preserve product qualities, unusual temperatures, pressure, and so on, must be maintained in the column. Unless any of these conditions prevail, a multiple-column adsorber is preferable because of the operating flexibility it provides.

Multiple-Column System

The choice of a multiple-column system is applicable when the nature of the process does not allow for interruption during loading, unloading, or regeneration especially when an alternate unit is not available. Multiple systems are also preferable when space constraints do not allow for a single column of adequate height or residence time.

The layouts of various fixed-bed adsorption systems and parallel fixed-bed adsorption systems are illustrated in Chapter 2. The columns of a parallel-column system are onstream at even time intervals and the column discharge is to a common manifold. The parallel design allows for smaller pumps, lower power requirements, and less stringent pressure specifications for columns and piping. Normally the carbon is not completely spent at the point where it is removed for regeneration.

The effluent from one column of a series-column system becomes the feed for the next column. The series layout is preferred over the parallel layout if the highest possible effluent purity is desired and the breakthrough curve is gradual or if the combination of a gradual breakthrough curve and high carbon demand per unit of production economically needs to exhaust the carbon. The carbon in the lead column is removed during regeneration and new carbon is put in the column onstream at the end of the series. The former lead column is replaced by the second column. The result is that normally the operating costs for series systems are lower than for single or parallel column in the same application. Where the two layouts are combined in a series and parallel system, the best characteristics of each layout are realized.

The fixed-bed systems described can either have upward or downward liquid flows. Downflow operation has more of an inherent filtering capability. Suspended solids will be removed by the finer carbon particles at the top of the bed. The capture of significant quantities of suspended solids can lead to high-pressure drops. At this point the procedure is to backwash the adsorber which can take time and use significant wash liquid. Therefore, the downflow operation must have piping in both directions. The direction of flow is the same during adsorption and washing cycles for the upflow operations. The washing cycles are far less frequent, pressures drops are lower, and considerably less downtime and wash liquid are consumed. Although some filtration will occur, upflow operations will not produce an effluent free of turbidity or suspended solids. Applications are further discussed in Chapter 2.

Pulsed-Bed Adsorbers

The carbon moves countercurrent to the liquid in pulsed-bed adsorbers. The effect is of a number of stacked, fixed-bed columns operating in series. Spent carbon is removed from the bottom of the columns as the liquid flows upward and fresh or segmented carbon is supplemented at the top. Pulsed-bed columns are usually operated with the columns completely filled with carbon which does not allow for bed expansion during operation or cleaning. Where the pulsed-bed unit does permit bed expansion, the efficiency of the unit deteriorates due to the mixing carbon disturbing the adsorption zone. The withdrawn carbon may contain spent and partially spent carbon.

Pulsed-bed adsorbers are most commonly operated on a semicontinuous basis. During this type of operation, a set quantity of spent carbon is removed at defined intervals from the bottom of the column. Replacement of carbon is at the top. Pulsed-bed systems are the type of liquid-phase carbon adsorption which comes closest to completely exhausting the carbon with the least capital investment.

The goal of either system (pulsed or fixed bed) is to maximize the use of the carbon by regenerating just the carbon that is expended. The choice of a pulsed-bed system is generally made when the feed does not contain suspended solids and the usage rate for carbon is high. Pulsed-bed systems are not effective for biologically active feeds. Conversely, fixed-bed systems are normally employed when the liquid contains significant quantities of suspended solids, is biologically active, or carbon usage is low.

Regeneration

Various alternative regeneration techniques for restoring spent carbon to its original adsorptive capacity make use of thermal, biological, chemical, hot-gas, or solvent techniques. Multihearth or rotary furnaces can be used to volatilize and carbonize adsorbed materials. Aerobic, anaerobic, or both types of bacteria can be used on site to remove adsorbed biodegradable material. Some methods are destructive to the adsorbate and no recovery can be made. Chemical, hot-gas, steam, and solvent regeneration are nondestructive methods for recovery of materials. All are carried out in place and rely on the varying adsorptive capacity of carbon for organics under changing process conditions such as pH, temperature, and nature of the liquid phase. Chemical regeneration uses a regenerant such as formaldehyde to react with the sorbed material and remove it from the carbon. Hot-gas regeneration is used when carbon has adsorbed a low boiling point organic material. Steam, CO_2, or M_2 are passed through the bed causing vaporization. Solvent regeneration employs a suitable solvent to pass through the spent carbon and dissolve the adsorbed material. Solvents are then recycled and purified, usually through decantation or batch distillation. Chemical and solvent regeneration methods can be combined effectively. Steam is widely used for low-temperature regeneration.

Labor Requirements

For liquid-phase carbon adsorption, labor requirements range from the more labor-intensive role of a batch-sequence operator to the more continuous liquid-phase adsorption. Batch-sequence operators typically perform the following sequence of operations. The adsorption vessel is charged with the liquid process stream. Activated carbon is added in variety of possible ways. If charging is automated or dust controlled, the operator runs this equipment. The operator typically runs the agitator for a set time cycle, then activates a pump to draw the treated mixture through a filter to remove the carbon. The attention required during the filtration cycle is highly dependent on the size of the operation and the filter capacity. After filtering, the operator washes and dries the cake as needed and removes it from the vessel. For small batch operations, the filter-cake removal operation is manual in most cases. The operator then packages the carbon in an acceptable manner for disposal in nonregenerable operations. In a regenerable operation the operator may have the responsibility of charging the spent carbon to the regeneration operation, such as a multihearth or rotary furnace in the case of heat regeneration.

For a continuous liquid-phase adsorption system, less direct operator presence than for batch adsorption is required. More of the sequences are automatically controlled and the continuous nature of the operation allows a supervisory role for the operator. Where the adsorption chambers are periodically emptied and recharged with fresh carbon on a regular basis, additional labor is needed. The actual labor requirement of adsorption operations is a function of the total system design which includes all associated processing facilities. For both batch and continuous operations it is, to an extent, determined by the magnitude of the operation (number of units, size of equipment, complexity of operation) and the nature of the total system design which could, for example, range from a small pharmaceutical production to a large sugar refinery operation. On a case-by-case basis, certain decisions must be made as to whether the adsorption system alone is being considered in analyzing labor needs or whether associated procedures (that is, filtration) should be considered.

INDUSTRIAL APPLICATIONS

The following are typical industrial applications for liquid-phase carbon adsorption and are further detailed in Chapter 2. Generally liquid-phase carbon adsorbents are used to decolorize or purify liquids, solutions, and liquefiable materials such as waxes. Specific industrial applications include the decolorization of sugar syrups; the removal of sulfurous, phenolic, and hydrocarbon contaminants from wastewater; the purification of various aqueous solutions of acids, alkalies, amines, glycols, salts, gelatin, vinegar, fruit juices, pectin, glycerol, and alcoholic spirits; dechlorination; the removal of grease from dry-cleaning solvents and from electroplating solutions; and the removal of wastes, aniline, benzene, phenol, and camphor from water. The use of adsorption for the removal of trace contaminants, as in the commercial use for the recovery of major components of feed streams as pure products, has been a widespread application.

GASEOUS-PHASE ADSORPTION

Gaseous-phase carbon adsorption systems can be classed in several ways. The first category is between regenerable and nonregenerable processes. The majority of industrial systems are regenerable operations that allow the user to recover the adsorbate and continue to reuse the activated carbon adsorbent. Regeneration relies on the continuity of gaseous adsorption achieved through equipment cycling to a desorption or regeneration phase of operation in which the temporarily exhausted beds of carbon are generated by removing the adsorbate. Regeneration operations are categorized in the following mechanisms: thermal swing regeneration, pressure swing regeneration, inert gas purge stripping, and displacement desorption.

Thermal swing is widely used for regeneration in purification adsorption operations. The spent bed is heated to a level at which the adsorptive capacity is reduced so that the adsorbate leaves the activated carbon surface and is removed in a stream of purge gas. *Pressure swing* relies on the reduction of pressure at constant temperature to reduce the adsorptive capacity for an adsorbate. Pressures can drop from elevated to atmospheric or from atmospheric to vacuum conditions. *Inert purge stripping* relies on the passage of a liquid or gas, without adsorbable molecules and in which the adsorbate is soluble, through the spent carbon bed at constant temperature and pressure. *Displacement desorption* relies on the passage of a fluid containing a high concentration of an adsorbable molecule or a more strongly adsorbable molecule than the adsorbate presently on the carbon.

Gaseous-phase adsorption systems are also categorized as either fixed-bed adsorbers or movable-bed adsorbers.

Fixed-Bed Adsorbers

The various configurations of fixed-bed gaseous-phase carbon adsorption systems are illustrated in various sections throughout this book as well as in Figures 1.1, 1.2, and 1.3. Enclosures for simple fixed-bed adsorbers may be vertical or horizontal, cylindrical or conical shells. Where multiple fixed-beds are needed, the usual configuration is a vertical cylindrical shell. The type of enclosure used is normally dependent on the gas volume handled and the permissible pressure drop.

The gas flow can be either down or up. Downflow allows for the use of higher gas velocities, while in upflow the gas velocity must be maintained below the value which prevents carbon boiling which damages the bed. When large volumes of gas need to be handled, cylindrical horizontal vessels are selected. The beds are oriented parallel to the axis.

For the continuous operation of fixed-bed adsorbers, it is desirable to have two or three units. With two adsorber units one unit adsorbs while the other regenerates or desorbs. The required times for regeneration and cooling of the adsorbent are the factors determining the cycle time. Under most situations, two adsorbing units are sufficient if the regeneration and cooling of the second bed can be completed prior to the breakthrough of the first unit. The move to three units makes it possible for one bed to be adsorbing, one cooling, and the third regenerating. The vapor-free air from the first bed is used to cool the unit which was just regenerated. Occasionally a fourth bed is used. An example arrangement of four beds would be to have two units

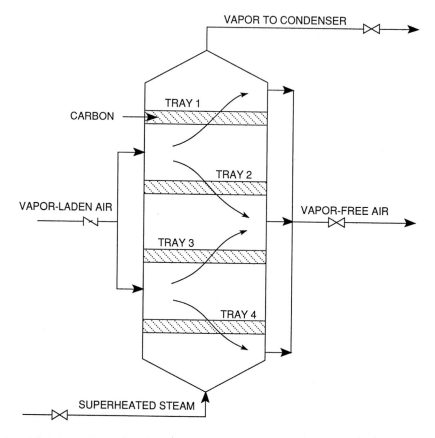

Figure 1.1 Cross section of adsorber with four fixed beds of activated carbon.

adsorbing in parallel, discharging exhaust to a third unit on the cooling cycle as a fourth unit is regenerated. Figure 1.4 shows a diagrammatic sketch of a two-unit fixed-bed adsorber. A three-bed adsorber configuration is illustrated in Figure 1.5.

Conical fixed-bed adsorbers are employed when a low-pressure drop is desired. For systems of the same diameter and same weight of carbon, the

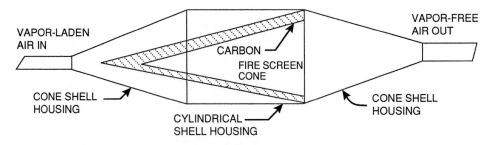

Figure 1.2 Horizontal adsorber.

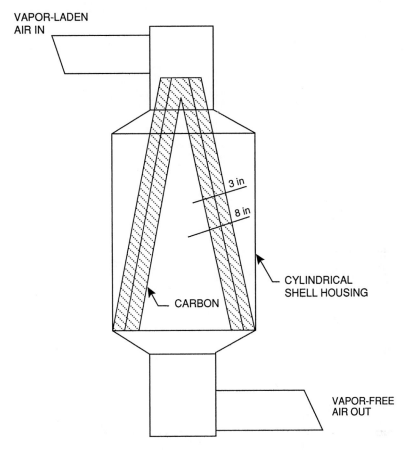

Figure 1.3 Diagrammatic sketch of a vertical adsorber with two cones, permitting studies on different depths of carbon beds.

pressure drop through the cone-shaped bed is less than one half of that through a conventional flat-bed adsorber, while the air volume is more than double that through the flat bed.

Movable-Bed Adsorbers

These are primarily used for solvent recovery and consist of a totally enclosed rotating drum housing which encloses a bed of activated carbon. A fan delivers solvent-laden air into the enclosure through ports and into the carbon section above the bed. The solvent-laden air passes through the bed to a space on the inside of the cylindrical carbon bed. The clean-air discharge is through ports at the end of the drum opposite the entrance, axially to the drum and out to the atmosphere. Steam is normally used to regenerate the movable-bed adsorbers. A continuous carbon adsorber is diagrammed in Figure 1.6.

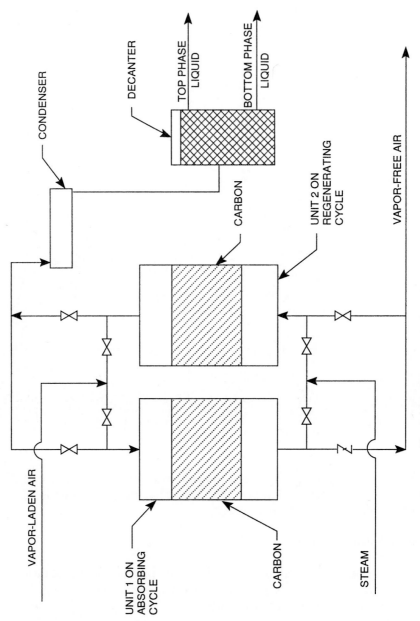

CONDENSER

DECANTER

TOP PHASE LIQUID

BOTTOM PHASE LIQUID

CARBON

UNIT 2 ON REGENERATING CYCLE

VAPOR-FREE AIR

VAPOR-LADEN AIR

UNIT 1 ON ABSORBING CYCLE

CARBON

STEAM

Figure 1.4 Diagram of a two-unit, fixed-bed adsorber.

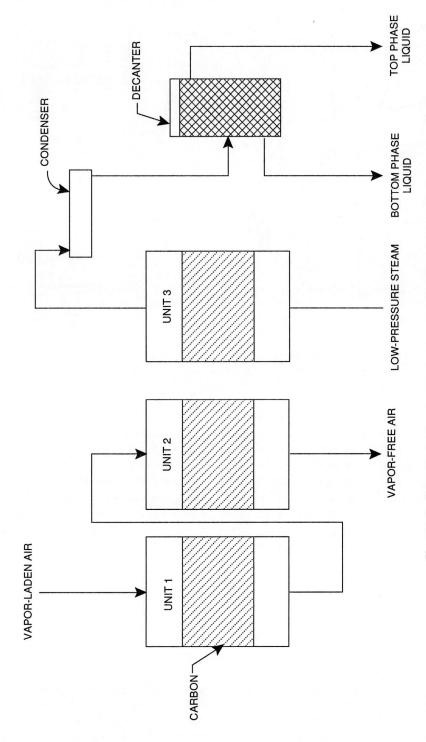

Figure 1.5 Diagram of a three-unit operation of a fixed-bed adsorber showing No. 1 and No. 2 adsorbing in a series and No. 3 regenerating. Second cycle, No. 2 and No. 3 will be adsorbing with No. 1 regenerating. Final cycle, No. 3 and No. 1 will be adsorbing with No. 2 regenerating.

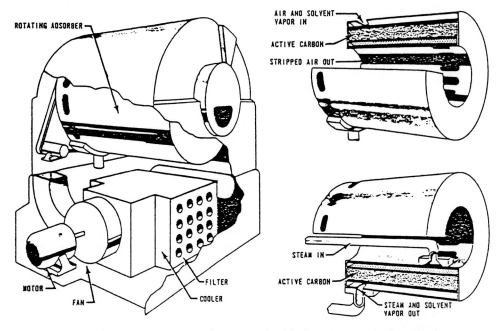

Figure 1.6 Left: Diagram of a rotating fixed-bed continuous adsorber showing the path of the vapor-laden air to the carbon bed. Right: Cut of continuous adsorber showing path of steam during regeneration.

In a typical start-up of a gas adsorption system, the operator conducts a general check of all system components: gaskets, bypass valves, adsorber and alternate unit, temperatures. Time clocks for the desired cycle times are set controlling adsorption times, purge time, hot-gas drying, and cooling. Following the start of gas flow, pressure drops and temperature rises are monitored. In the case of multiple units, the operator checks to see that desorption is proceeding properly; the stripping fluid is turned on; the regeneration or disposal system for the dissolved vapors is checked; the condenser cooling-water lines are opened and the cooling-water exit temperature is monitored.

During operation, temperatures and pressure drops are monitored. The prefilter is monitored to insure adequate gas flow to the processing system and is changed when a high-pressure drop is recorded. A check is made on the bed to be regenerated to make sure the proper carbon bed sequencing is set and the bed is prepared for the steam or stripping medium. When a third bed is used, a check is made to insure that this is being properly cooled. There should be a provision for routinely monitoring the adsorber emissions. Although a proper cycling procedure has been used between alternate beds, an unexpected contamination of the adsorbent would cause a premature breakthrough of the beds resulting in the release of contaminants. A routine shutdown would normally only involve the shut off of the gas flow from the process. If the adsorber were to be shut down for a lengthy period of time, complete draining of all lines and vessels normally would be practical.

The maintenance duties are relatively simple. Frequent inspection for equipment abrasion and corrosion is part of a preventive maintenance program. Filters need to be inspected frequently to check for holes or plugging and when necessary they are cleaned and/or replaced. Other maintenance duties consist of maintaining adequate oil levels in lubrication reservoirs; inspecting the general condition of the exterior; periodic testing for bed contaminants; inspecting/repairing of pumps blowers, valves, and so on; inspecting recovery system controls including cooling-water temperature, condenser temperatures, and so on; inspecting built-in safety devices; and inspecting the regeneration system. The frequency of the preventive maintenance varies from daily to monthly depending on the application and the manufacturer's suggestions.

Industrial applications for gaseous-phase carbon adsorption have been discussed. The major use of carbon adsorption for gaseous systems is in solvent recovery from exhaust leaving a process evaporation chamber. This type of use is employed in graphic arts operations, various types of spray-painting applications, textile dry-cleaning operations, and polymer processing. Carbon adsorption is used extensively in the control of vapors from petroleum marketing plants. While regulations limit the emission of volatile organic compounds (VOCs), most of these gas adsorption operations would be practical anyway because of the value of the recovered solvent. Less common applications include the fractionation of gases, low molecular weight hydrocarbons, rare gases, industrial gases and the purification of intake gases, circulating air, or process exhaust air to remove odors, toxic gases, and so on.

ADSORPTION THEORY

When brought into contact with a solid substance, a gas or vapor has the tendency to collect on the surface of the solid. This phenomenon is known as *adsorption*. The amount of adsorption on the surface of most solids is exceedingly small but materials such as activated alumina, silica gel, and activated carbon have been developed to adsorb gases and vapors on their surfaces. These materials are porous solids which have an unusually high surface development in the form of an ultramicroporous structure; thus, these materials possess a very large internal surface. A fluid is able to penetrate through the pore structure of these materials and be in contact with the large surface area available for adsorption.

The mechanism by which this surface adsorption takes place is complex. Many theories have been offered to explain adsorption, details of which may be found in the literature. The important types of adsorption are physical adsorption, in which case the gas is attracted to the surface of the adsorbent, and chemical adsorption in which the gas shows a strong interaction with the manner of a chemical reaction. The surface attraction is due to van der Waals forces: the intermolecular forces that produce normal condensation to the liquid state. On smooth surfaces the van der Waals adsorption is restricted to a layer of not more than a very few molecules in thickness. But in a porous solid with a capillary structure, the surface adsorption is supplemented by capillary condensation which is also brought about by the van der Waals forces of attraction.

Adsorptive processes are exothermic. The heat of adsorption due to surface attraction is greater than the heat of condensation of the gases being adsorbed. The heat of chemical adsorption increases progressively with increases in the partial pressure of the gas. At low partial pressure, adsorption is by surface attraction, and at higher partial pressures the smallest wetted capillaries become effective and condensation begins. At the higher pressure the larger capillaries become effective.

Figure 1.7 shows the relationship between maximum effective pore size and vapor pressure (or vapor concentration) for benzene at 20°C. These sizes are computed on the basis of capillary condensation theory. The relationship between the pressure and the amount adsorbed is dependent on the size distribution of the capillary pores as well as the area of the exposed surface and the nature of the adsorbent and the gas.

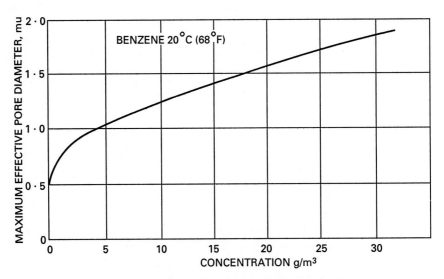

Figure 1.7   Relationship between pore size and vapor concentration.

These relationships can be expressed graphically in the form of adsorption isotherms. Determined experimentally, these relationships are expressions of the amount of gas adsorbed under true static equilibrium conditions.

ADSORPTION ISOTHERMS

Various types of isotherms are observed and the shapes of the graphs vary according to the adsorption system. According to Brunauer, five types of adsorption isotherms exist in the literature on the adsorption of gases, and these are shown in Figure 1.8. Adsorption is specific depending on the nature of the system. Preferential adsorption is of importance for the selective removal of compounds from fluid mixtures. When mixtures are adsorbed, the presence of each affects the equilibrium of the others. In general, a molecule of high molecular weight, high critical temperatures, and low volatility is adsorbed in preference to one of low molecular weight, low critical temperature, and high volatility. A preferentially adsorbed molecule will displace others which have already been adsorbed.

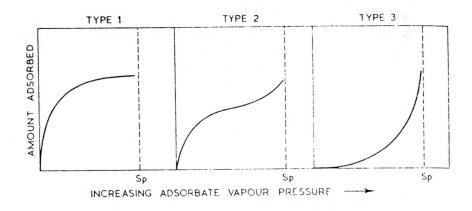

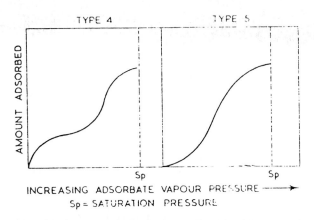

Figure 1.8 Types of adsorption isotherms: type 1, adsorption of oxygen on carbon at −183°C (−298°F); type 2, nitrogen on iron catalyst at −195°C (−319°F); type 3, bromine on silica gel; type 4, benzene on ferric oxide gel at 50°C (122°F); type 5, water vapor on carbon at 100°C (212°F).

Use of this preferential adsorptive property is made in selective adsorption by actuated carbon of a single hydrocarbon gas from a mixture. In such applications the activated carbon may be regarded as a rectification plant. By adjustment of the adsorption period, any particular cut can be produced from the fractions contained in the gas. For example, in a hydrocarbon mixture, methane which is adsorbed first will be displaced by ethane which in turn will be displaced by the propane, and this process of displacement will continue throughout the series of hydrocarbons.

DESORPTION OR REGENERATION

Regeneration, desorption, or stripping of the adsorbed gases from an adsorbent may be accomplished in a number of ways:
  - The temperature of the adsorbent may be raised until the vapor pressure of the adsorbed gas exceeds atmospheric pressure. The adsorbed gas will then be evolved and may be collected at atmospheric pressure.
  - Adsorbed gas may be withdrawn by vacuum application lowering the pressure below the vapor pressure of the adsorbed gas.
  - Adsorbed gas may be withdrawn in a stream of inert gas passing through the adsorbent, keeping the partial pressure of the stripped gas in the gas stream below the equilibrium pressure of the adsorbed gas.
  - Adsorbed gas may be withdrawn in a stream of an easily condensable gas such as steam. Stripped gas is recovered by condensing the stripped gas and steam mixture.
  - Adsorbed gas may be displaced by the adsorption of a gas which is preferentially adsorbed.

MANUFACTURE OF ACTIVATED CARBON

Activated carbon may be manufactured from a wide range of carbonaceous substances including bones, coals, wood dust, peat, nutshells, and wood charcoal. The fine capillary structure is formed during the activation process. The raw material is a compound of amorphous carbon and hydrocarbons not active but becoming active when the hydrocarbons held in the carbon are removed by oxidation. The combined effect produces an ultrafine capillary structure throughout the carbon. The principal commercial methods used are chemical activation and steam activation.

The chemical method of activation consists of mixing a pulverized form of carbonaceous material with a liquid dehydrating agent, drying the mixture, and heating in a retort to complete the activation. The most widely used chemicals are zinc chloride and phosphoric acid.

Raw material for steam activation is usually a carbonized material derived from wood, peat, brown coal, and so on. The charcoal is heated in a retort to a high temperature. Steam, the activating agent, is passed through the bed of heated charcoal to produce the desired porous structure, and the resulting water gas produced is often used for heating the retorts.

The quality and characteristics of activated carbon depend on the physical properties of the raw materials and methods of activation used. There is a wide choice of raw materials and control of the activation process, producing activated carbons having widely varying physical and adsorptive properties.

Differing physical properties of the many types and grades of commercially available activated carbon make selection difficult. It is therefore advisable to discuss the application with the carbon manufacturer in order to obtain the carbon grade best suited for the intended application. Most manufacturers are willing to cooperate and offer useful information regarding the application of their materials.

PROPERTIES OF ACTIVATED CARBON

The great advantage that activated carbon has over other recovery systems is the outstanding ability to recover organics from low concentrations easily and inexpensively. This is desirable in processes using flammable or toxic concentrations easily and inexpensively. This is desirable in processes using flammable or toxic compounds, where it is necessary for safety reasons to ensure adequate ventilation to prevent solvent concentrations from reaching dangerous proportions.

Gas purification applications discussed in Chapters 2 and 4 involving the removal of small concentrations of impurities include the deodorization of air, the removal of traces of organic impurities from gas streams to prevent catalyst poisoning, the removal of traces of oil vapors from compressed gases, and the removal of similar substances from fluid streams. Important gas separations using activated carbon are discussed in subsequent chapters.

OTHER ADSORBENTS

Silica Gel

Silica gel is a granular adsorbent having a translucent appearance. Manufacture consists essentially of adding a solution of sodium silicate to sulfuric acid, washing the gel with alcohol, drying, roasting, and grading. Silica gel shows a specific selective adsorption of water vapor which is higher than either activated carbon or activated alumina and, hence, its principal application is for the dehydration of gases. It is capable of adsorbing up to 40 percent of its weight of water vapor and may be simply reactivated by passing heated gas or air through the adsorbent and cooling.

Activated Alumina

Aluminum oxide base adsorbents are prepared by heat treatment of bauxite or alumina hydrate, producing a porous solid adsorbent. Activated alumina has the higher adsorbent capacity but because of its lower cost, activated bauxite has found desiccant applications in competition with silica gel and activated alumina. Both activated alumina and activated bauxite are also widely used for the dehydration of gases. Reactivation is carried out by passing heated gas or air through the spent adsorbent and cooling.

Preferential adsorption characteristics and physical properties of the industrial adsorbents determine the main applications for each type. All adsorbents are capable of adsorbing organic solvents, impurities, and water vapor from gas streams, but each has a particular affinity for water vapor or organic vapors.

Activated alumina, silica gel, and molecular sieves will preferentially adsorb water from a gas mixture containing water vapor and organic solvent. This is a serious disadvantage in solvent recovery work where the water content of the air or gas stream is often greater than the solvent content.

Silica gel and activated alumina disintegrate under liquid water, rendering their use for the desorption of organic solvents much more difficult. Such adsorbents, therefore, are normally used for the drying of air and gases and are regenerated by blowing a stream of hot air or gas through the adsorbent bed. Carbon is normally regenerated by direct steam to facilitate the collection of the stripped solvents by simply condensing the steam-solvent vapors produced.

2

Carbon Adsorption Applications

INTRODUCTION

Carbon has been known throughout history as an adsorbent with its usage dating back centuries before Christ. Ancient Hindus filtered their water with charcoal. In the thirteenth century, carbon materials were used in a process to purify sugar solutions. In the eighteenth century, Scheel discovered the gas adsorptive capabilities of carbon and Lowitz noted its ability to remove colors from liquids. Carbon adsorbents have been subjected to much research resulting in numerous development techniques and applications.

One of these applications was begun in England in the mid-nineteenth century with the treatment of drinking waters for the removal of odors and tastes. From these beginnings, water and wastewater treatment with carbon has become widespread in municipal and industrial processes, including wineries and breweries, paper and pulp, pharmaceutical, food, petroleum and petrochemical, and other establishments of water usage. Interest in carbon use for air as well as water pollution control and traditional industrial/product applications has received increased attention since the early 1970s with the advent of more stringent environmental regulations.

ADSORPTION PROCESS

The adsorption process occurs at solid-solid, gas-solid, gas-liquid, liquid-liquid, or liquid-solid interfaces. Adsorption with a solid such as carbon depends on the surface area of the solid. Thus, carbon treatment of water involves the liquid-solid interface. The liquid-solid adsorption is similar to the other adsorption mechanisms. There are two methods of adsorption: physisorption and chemisorption. Both methods take place when the molecules in the liquid phase become attached to the surface of the solid as a result of the attractive forces at the solid surface (adsorbent), overcoming the kinetic energy of the liquid contaminant (adsorbate) molecules.

Physisorption occurs when, as a result of energy differences and/or electrical attractive forces (weak van der Waals forces), the adsorbate molecules become physically fastened to the adsorbent molecules. This type of adsorption is multilayered; that is, each molecular layer forms on top of the previous layer with the number of layers being proportional to the contaminant concentration. More molecular layers form with higher concentrations of contaminant in solution.

When a chemical compound is produced by the reaction between the adsorbed molecule and the adsorbent, chemisorption occurs. Unlike physisorption, this process is one molecule thick and irreversible because energy is required to form the new chemical compound at the surface of the adsorbent, and energy would be necessary to reverse the process. The reversibility of physisorption is dependent on the strength of attractive forces between adsorbate and adsorbent. If these forces are weak, desorption is readily effected.

Factors affecting adsorption include:

- The physical and chemical characteristics of the adsorbent, that is, surface area, pore size, chemical composition, and so on.
- The physical and chemical characteristics of the adsorbate, that is, molecular size, molecular polarity, chemical composition, and so on; the concentration of the adsorbate in the liquid phase (solution).
- The characteristics of the liquid phase, that is, pH and temperature.
- The residence time of the system.

ADSORPTION WITH ACTIVATED CARBON

Certain organic compounds in wastewaters are resistant to biological degradation and many others are toxic or nuisances (odor, taste, color forming), even at low concentrations. Low concentrations may not be readily removed by conventional treatment methods. Activated carbon has an affinity for organics and its use for organic contaminant removal from gaseous streams and wastewaters is widespread.

The effectiveness of activated carbon for the removal of organic compounds from fluids by adsorption is enhanced by its large surface area, a critical factor in the adsorption process. The surface area of activated carbon typically can range from 450–1,800 m^2/g. Some carbons have been known to have a surface area up to 2,500 m^2/g and examples are shown in Table 2.1.

TABLE 2.1   SURFACE AREAS OF
TYPICALLY AVAILABLE ACTIVATED
CARBONS

Origin	Surface area (m^2/g)
Bituminous coal	1,200–1,400
Bituminous coal	800–1,000
Coconut shell	1,100–1,150
Pulp mill residue	550–650
Pulp mill residue	1,050–1,100
Wood	700–1,400

Of less significance than the surface area is the chemical nature of the carbon's surface. This chemical nature or polarity varies with the carbon type and can influence attractive forces between molecules. Alkaline surfaces are characteristic of carbons of vegetable origins and this type of surface polarity affects adsorption of dyes, colors, and unsaturated organic compounds. Silica gel, an adsorptive media that is not a carbon compound, has a polar surface which also exhibits an adsorptive preference for unsaturated organic compounds as opposed to saturated compounds. However, for the most part, activated carbon surfaces are nonpolar, making the adsorption of inorganic electrolytes difficult and the adsorption of organics easily effected.

ACTIVATED CARBON CHARACTERISTICS

Pores of the activated carbon exist throughout the particle in a manner illustrated in Figure 2.1. The pore structure of activated carbon affects the large surface-to-size ratio. The macropores do not add appreciably to the surface area of the carbon but provide a passageway to the particle interior and the micropores. The micropores are developed primarily during carbon activation and result in the large surface areas for adsorption to occur.

Macropores are those pores greater than 1,000 A; micropores range between 10–1,000 A. Pore structure, like surface area, is a major factor in the adsorption process. Pore-size distribution determines the size distribution of molecules that can enter the carbon particle to be adsorbed. Figure 2.2 illustrates the discriminatory practices of the pores. Large molecules can block off micropores, rendering useless their available surface areas. However, because of irregular shapes and constant molecule movement, the smaller molecules usually can penetrate to the smaller capillaries.

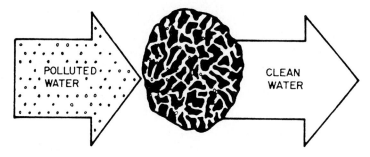

Figure 2.1 Artist's conception of a carbon granule. Organics along with the water pass through the pores and become adsorbed on the pore surfaces.

TABLE 2.2 PROPERTIES OF SEVERAL COMMERCIALLY AVAILABLE GRANULATED CARBONS[a]

	ICI America Hydrodarco 3000	Calgon Filtrasorb 300 (8 × 30)	Westvaco Nuchar W-L (8 × 30)	Witco 517 (12 × 30)
Physical Properties				
Surface Area, m²/gm (BET)	600–650	950–1,050	1,000	1,050
Apparent Density, g/cc	0.43	0.48	0.48	0.48
Density, Backwashed and Drained, lb/ft³	22	26	26	30
Real Density, g/cc	2.0	2.1	2.1	2.1
Particle Density, g/cc	1.4–1.5	1.3–1.4	1.4	0.92
Effective Size, mm	0.8–0.9	0.8–0.9	0.85–1.05	0.89
Uniformity Coefficient	1.7	≤1.9	≤1.8	1.44
Pore volume, cc/g	0.95	0.85	0.85	0.60
Mean Particle Diameter, mm	1.6	1.5–1.7	1.5–1.7	1.2
Specifications				
Sieve Size (U.S. standard series)				
Larger than No. 8, max. %	8	8	8	[b]
Larger than No. 12, max. %	[b]	[b]	[b]	5
Smaller than No. 30, max. %	5	5	5	5
Smaller than No. 40, max. %	[b]	[b]	[b]	[b]
Iodine No.	650	900	950	1,000
Abrasion No., minimum	[c]	70	70	85
Ash, %	[c]	8	7.5	0.5
Moisture as packed, max. %	[c]	2	2	1

[a] Other sizes of carbon are available on request from the manufacturers.
[b] Not applicable to this size carbon.
[c] No available data from the manufacturer.

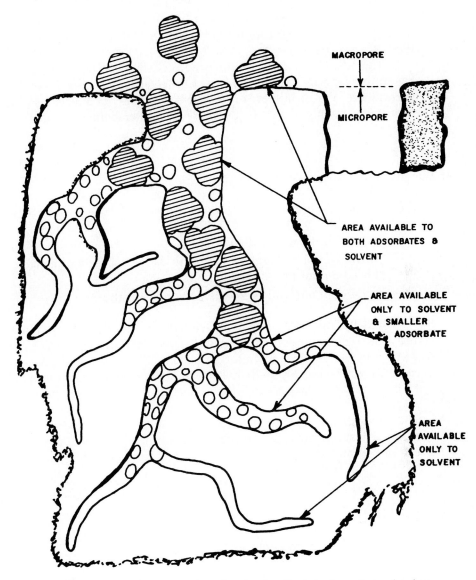

MACROPORE

MICROPORE

AREA AVAILABLE TO
BOTH ADSORBATES &
SOLVENT

AREA AVAILABLE
ONLY TO SOLVENT
& SMALLER
ADSORBATE

AREA
AVAILABLE
ONLY TO
SOLVENT

Figure 2.2 Artist's conception of molecular discrimination effects of carbon pores.

Since adsorption is possible only in those pores that can be entered by molecules, the carbon adsorption process is dependent on the physical characteristics of the activated carbon and the molecular size of the adsorbate. Each application for carbon treatment must be cognizant of the characteristics of the contaminant to be removed and designed with the proper carbon type in order to attain optimum results. Table 2.2 gives the properties of some commercially available granulated activated carbons.

Basically, there are two forms of activated carbon: powdered and granular. The former are particles that are less than U.S. Sieve Series No. 50, while the latter are larger.

The adsorption rate is influenced by carbon particle size, but not the adsorptive capacity which is related to the total surface area. By reducing the particle size, the surface area of a given weight is not affected. Particle size contributes mainly to a system's hydraulics, filterability, and handling characteristics.

CARBON ACTIVATION

Carbon materials are activated through a series of processes which includes:

- Removal of all water (dehydration).
- Conversion of the organic matter to elemental carbon, driving off the noncarbon portion (carbonization).
- Burning off tars and pore enlargement (activation).

Initially, the material to be converted is heated to 170°C to effect water removal. Temperatures are than raised above 170°C driving off CO_2, CO, and acetic acid vapors. At temperatures of about 275°C, the decomposition of the material results and tar, methanol, and other by-products are formed. Nearly 80 percent elemental carbon is then effected by prolonged exposure to 400–600°C.

Activation of this product follows with the use of steam or carbon dioxide as an activating agent. The superheated steam, 750–950°C, passes through the carbon burning out by-product blockages, and expanding and extending the pore network.

CARBON SYSTEMS

Water, Wastewater, and Activated Carbon

Activated carbon is commonly used in water and wastewater treatment, removing organics that cause odors, tastes, and other detrimental effects. In addition, as a recycling media, activated carbon can be used for solvent purification or recovery of expensive materials. The economics of carbon systems have been improving such that their usage is becoming more accepted.

The utilization of carbon at a wastewater or water treatment facility can be in the form of a powder or granule. Granular carbon is placed in a bed and raw water or wastewater is passed over it. Tastes, colors, and odors are removed from potable waters and dissolved organics, such as phenols, pesticides, organic dyes, surfactants, and so on, are removed from industrial and municipal wastewaters. Table 2.3 gives applications for activated carbon's adsorption abilities for various compounds.

TABLE 2.3   SEVERAL ACTIVATED CARBON REMOVAL APPLICATIONS

Acetaldehyde	Gasoline
Acetic Acid	Glycol
Acetone	Herbicides
Activated Sludge Effluent	Hydrogen Sulfide
Air Purification Scrubbing Solutions	Hypochlorous Acid
Alcohol	Insecticides
Amines	Iodine
Ammonia	Isopropyl Acetate and Alcohol
Amyl Acetate and Alcohol	Ketones
Antifreeze	Lactic Acid
Benzine	Mercaptans
Biochemical Agents	Methyl Acetate and Alcohol
Bleach Solutions	Methyl-Ethyl-Ketone
Butyl Acetate and Alcohol	Naptha
Calcium Hypochlorite	Nitrobenzenes
Can and Drum Washing	Nitrotoluene
Chemical Tank Wash Water	Odors
Chloral	Organic Compounds
Chloramine	Phenol
Chlorobenzene	Potassium Permanganate
Chlorine	Sodium Hypochlorite
Chlorophenol	Solvents
Chlorophyl	Sulfonated Oils
Cresol	Tastes (Organic)
Dairy Process Wash Water	Toluene
Decayed Organic Matter	Trichlorethylene
Defoliants	Trickling Filter Effluent
Detergents	Turpentine
Dissolved Oil	Vinegar
Dyes	Well Water
Ethyl Acetate and Alcohol	Xylene

The removal process continues until the carbon reaches its adsorption saturation point, at which time it is regenerated. The recoverable and waste products are extracted with a regeneration solution, the former being reused and the waste discharged. Figure 2.3 is a simple schematic of a granular activated carbon-bed system.

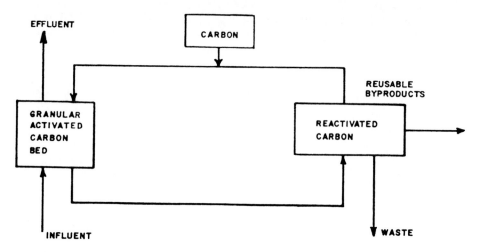

Figure 2.3 Simple schematic of a granular activated carbon-bed system.

Carbon Treatment Techniques

Powdered activated carbon is usually added to the water or wastewater stream with automatic chemical feeders. These feeders meter in the carbon at a predetermined rate at various points in the system. The point of carbon application varies with the treatment process and the desired results.

Carbon added during the early stages of treatment effects a more stable sludge and a better floc formation. In addition, because the carbon has adsorbed much of the organic matter, less chlorine disinfectant is required. Should carbon costs be high, it would be advantageous to administer the carbon later in the process stream. After sedimentation, less carbon would be required because an appreciable amount of the organic matter is removed in the flocculation process. Tertiary treatment involves the addition of carbon after filtration as a final polishing process. Best efficiencies with carbon are usually obtained with several points of application.

Granulated activated carbon beds are used in a similar fashion. Like powdered carbon, carbon-bed unit processes can be implemented at various points of a treatment plant. A carbon bed, situated just after the primary treatment and chemical addition process, would not only be used to remove dissolved organics, but also biodegradable organics, suspended solids, and colloidal materials. This is known as the physical-chemical treatment (PCT) process and does not involve biological treatment, that is, activated sludge, trickling filters, and biodisks.

PCT results in cost savings since the biological treatment unit is omitted. However, increased carbon loadings from PCT result in high regeneration rates (costs), thus minimizing the capital investment savings. Further, depending on the raw wastewater loadings, PCT may not be able to effect the required effluent characteristics. Figure 2.4 illustrates various PCT flow schematics.

Carbon beds have been used effectively in conjunction with biological treatment to obtain a high-quality effluent. In following unit operations of biodegradation and filtration, carbon-bed adsorption has several advantages, including (1) organics, BOD, and COD levels are reduced and require shorter carbon-bed retention times; (2) operating costs are reduced because regeneration of the carbon is minimized and suspended solids would not clog the bed; and (3) biological growth on the carbon and its associated problems would be kept to a minimum. Figure 2.5 shows points of application of carbon-bed unit processes after biological treatment.

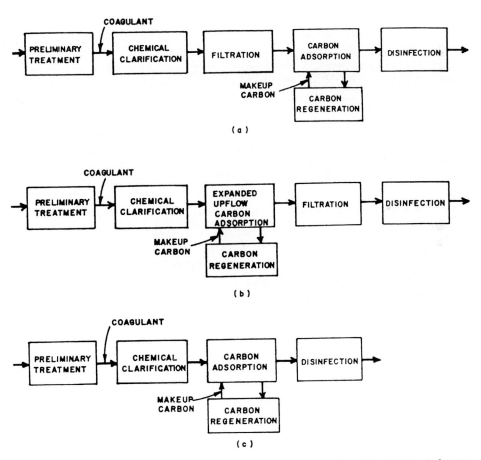

Figure 2.4 Various carbon application techniques in a physical-chemical treatment process.

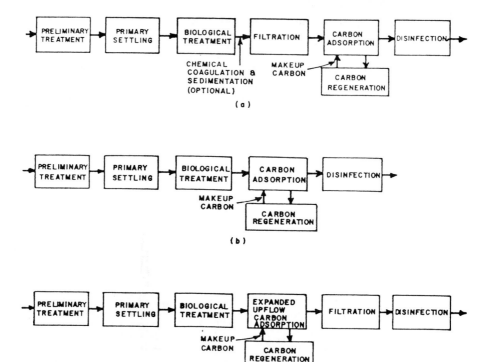

Figure 2.5 Points of application of carbon in a biological treatment plant.

The uses of activated carbon are many. Some of the advantages of an activated carbon system include:

- Considering the expense of industrial space, carbon beds require a relatively small amount of space.
- The process generates no secondary sludge.
- The process produces no odors. Recovery of valuable materials can be effected.

Wastewater streams, volume, and composition fluctuate; therefore, depending on the flow requirements and wastewater characteristics, an activated carbon system can be designed. Should the flow of wastewater at a facility vary drastically, a system such as the adsorption system, shown in Figure 2.6, could be employed. Since contact times and flow rates are dependent, a longer residence time can be obtained by reducing the flow rate through the system. High flow rates and/or long residence times can be accommodated with the addition of several module systems in series or parallel. This type of system is best suited for use at PCT-type facilities.

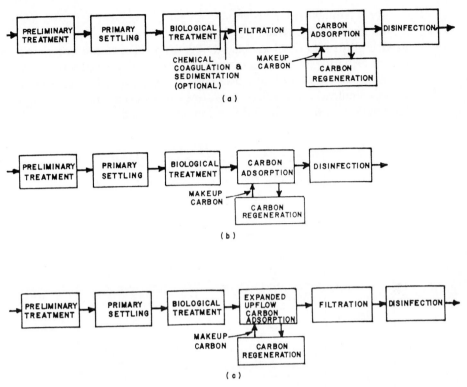

Figure 2.6 Application points for carbon in a bio-treatment plant where the pretreatment and primary settling portions can be used for equalization.

There are many types of carbon systems, each with its own advantages. Depending on flow rate, flow-rate variations, wastewater characteristics, effluent requirements, application, treatment process, and economics, a particular system can be selected. With each carbon process there are the inevitable trade-offs.

A countercurrent or upflow carbon bed effects a highly efficient use of carbon, reducing regeneration and carbon makeup costs. Further, it is a continuous operation that can be designed to eliminate downtimes for regeneration. However, this single-purpose use of carbon results in the need for other treatment units.

Cocurrent or downflow carbon beds not only can remove organics by adsorption, but also suspended solids by filtration. This results in a smaller capital investment because a filtration unit is no longer required. However, operating costs are increased as the carbon must be regenerated more frequently, and efficiencies are lower than if the carbon were used for a single purpose.

Another variable in a carbon-bed system would be the manner in which the wastewater is introduced to the carbon. The influent may flow through a carbon contactor due to gravity, in which case capital costs are kept to a minimum because conventional construction materials and techniques can be used. Pumps and their associated operating costs would not be required. On the other hand, with a pressurized flow through a carbon bed, a system has the capability of handling upset conditions more easily as flow and waste-water chemical composition variations can be accommodated. Figure 2.7 is a schematic of a pressurized downflow carbon bed.

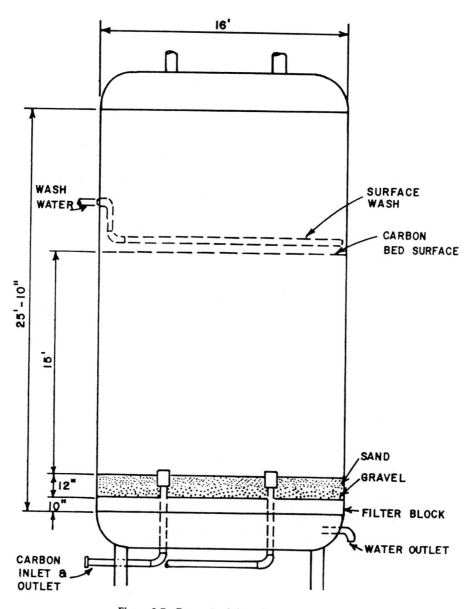

Figure 2.7   Pressurized downflow carbon bed.

A carbon bed's physical dimensions and carbon size as well as backwash requirements are subject to the containment loading of the wastewater and the flow rate. A high organically loading influent and flow rate would need a deep bed to effect a quality effluent. Generally, bed depths range from 10–30 ft and flow rates from 2–10 gpm/ft^2. A waste stream high in suspended solids content would have to be backwashed more frequently than one with a low suspended solid content. The suspended solids are filtered out of the waste stream by the carbon bed, which then begins to clog with the deposits.

Backwashing forces water through the bed in the opposite direction from normal, thereby removing the filtered debris from the carbon surfaces. Backwashing frequency may commonly be semiweekly up to a daily operation, should loadings be high.

Typical Arrangements

Columns of activated carbon are typically arranged in series to obtain a carbon-bed depth necessary to provide the required effluent quality. Wastewater flows through both (shown in Figure 2.8.) When the carbon in column No. 1 is spent, column No. 2 continues to remove contaminants to maintain water quality. Column No. 1 is taken off line and, after being flushed of spent carbon and refilled with new carbon, it is tied into the system downstream of column No. 2. The process is then repeated when column No. 2 becomes saturated

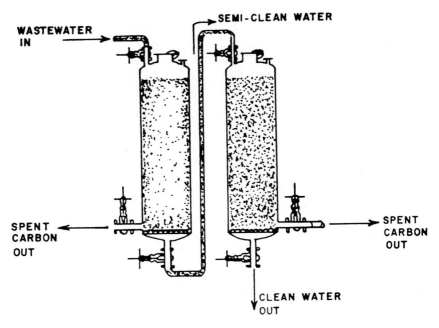

Figure 2.8 Multiple-column systems in series.

with impurities. Thus, this system can be used continuously with the carbon used most efficiently.

A typical column, as noted in Figure 2.7, is designed with flat or concave tops and bottoms. A retainer screen is placed on the bottom (filter block), followed by thin layers of gravel and sand (optional) and then the carbon. Common dimensions of a carbon vessel would involve height-to-diameter ratios of 2:1, but a higher ratio would improve a system's performance.

The release of the spent carbon from such vessels requires either a nozzle for hydraulic discharge or a man head for manual cleaning. These methods may necessitate backwashing or flushing to remove residual carbon.

To eliminate this problem of residual carbon left in the column during carbon replacement, cone-shaped bottoms are installed on the columns. A cone angle of 45° to 60° is normally used to facilitate adequate carbon flushing (Figure 2.9).

A moving-bed system is a modified upflow carbon bed and is generally used when large amounts of carbon are required for the removal of impurities. Like a countercurrent bed, wastewater flow is up through the carbon bed forcing the carbon particles apart and they expand in the bed. However, in a moving bed, the carbon also flows down through the column as spent carbon is periodically removed from the bottom. Fresh carbon is added at the top as the old is withdrawn to be regenerated.

Figure 2.10 illustrates this continuous-flow system. Depending on the operating parameters, this system can be highly efficient and economical. The

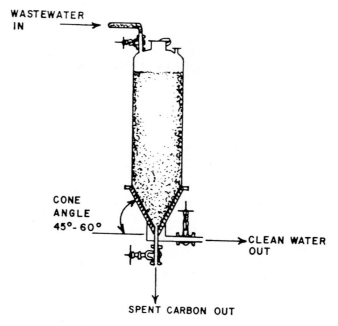

Figure 2.9   A cone-bottom column.

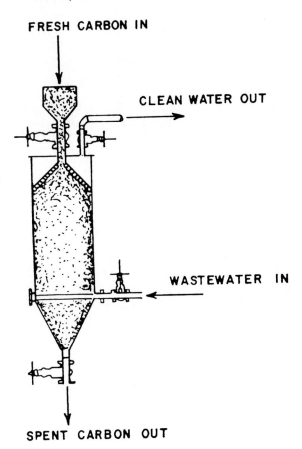

FRESH CARBON IN

CLEAN WATER OUT

WASTEWATER IN

SPENT CARBON OUT

Figure 2.10 Typical moving-bed system.

carbon removed from the bottom of this system has utilized nearly all its adsorption ability; whereas in a batch operation, the carbon efficiency is low. This system is also known as a pulsed carbon bed because the carbon is removed on an intermittent basis rather than continuously. Typical adsorber configurations are shown in Figure 2.11.

Carbon Regeneration

The economics of carbon are such that costs would be prohibitive if it could not be reused. Thus, spent carbon passes through a regeneration or reactivation process in which organics are desorbed and the carbon can be used again. This is usually a thermal process which proceeds as follows: The exhausted granulated carbon is withdrawn from the carbon-bed column and conveyed in the form of a water slurry. Before entering a rotary kiln or multihearth furnace, the slurry is dewatered. Furnace temperatures usually range between 1,600–1,800°F, during which time the carbon is dried of residual waters and the organic adsorbent is volatilized and oxidized. Combustion

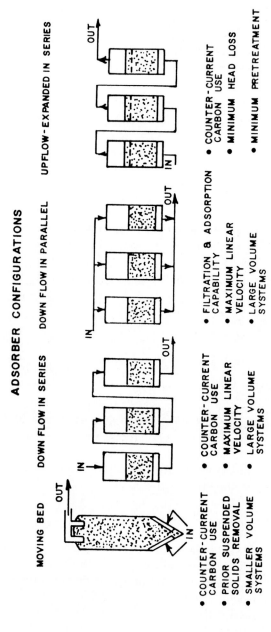

Figure 2.11   Typical carbon adsorption systems and their attributes.

34

conditions within the furnace are controlled to limit oxygen content to effect oxidation of the adsorbed material rather than the carbon.

After thermal regeneration, the carbon is quenched in a water bath, washed of carbon fines, and recycled back to the adsorber column, or sent to storage. Air pollution devices, such as scrubbers and afterburners, are installed on the furnace to control off-gas pollutants. Figure 2.12 is a schematic diagram of the process.

During each treatment cycle, carbon losses can vary between 2 percent and 10 percent. A carbon makeup hopper and bin are included in line to provide the additional carbon necessary for the purpose. The entire thermal regeneration process usually lasts about 30 minutes.

Regeneration of carbon without heat has been effected, but usually with a recovered adsorptive capacity of low levels. However, the cost disadvantages of low adsorptive capacity recovery level may be offset via by-product recovery.

Valuable chemicals can be recovered from a waste stream by passing them through a carbon column and then regenerating the carbon to effect product removal. An example of this is chromium. Chromium solution is introduced to a granular activated carbon column and the chromium is absorbed by the carbon. When the carbon's adsorptive capacity is exhausted,

GRANULAR CARBON REACTIVATION CYCLE

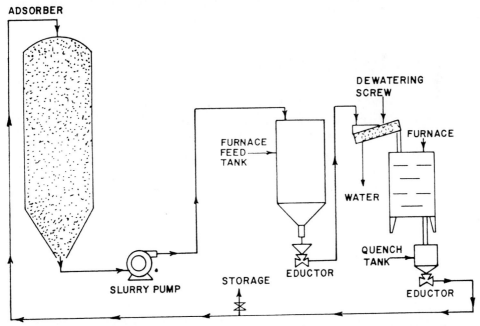

Figure 2.12 Schematic diagram of a granulated activated carbon thermal regeneration process.

it is regenerated. If the carbon is to be recycled, the carbon is regenerated for chromium recovery with a sodium hydroxide solution. Should the recovered adsorptive capacity of the carbon be so low that reuse would be impractical, then the chromium could be extracted from the carbon with a sulfuric acid solution. The carbon is then discarded.

The recovery of certain compounds by carbon adsorption can prove to be economically unfeasible due to high makeup carbon costs and a recovered product that may require further processing to effect purity.

Carbon Evaluation

To obtain the optimum advantage from activated carbon use, experimental analysis utilizing actual operating conditions is often necessary. The operating parameters of wastewater characteristics, treatment facility configuration, and effluent requirements are controlling factors in the selection of carbon type and mode of application.

The following list of characteristics can be evaluated for each activated carbon to determine its suitability for a particular application:

- *Surface area.* Generally, the larger the surface area, the more adsorption can take place.
- *Apparent density.* A measure of the regenerability of a carbon.
- *Bulk density.* Used to determine carbon quantities necessary to accomplish certain jobs.
- *Effective size, mean particle diameter, uniformity coefficient.* Used to determine hydraulic conditions of an adsorber column.
- *Pore volume.* Can be used to determine the adsorbability of a particular waste entity.
- *Sieve analysis.* Used to check plant-handling effects on the carbon.
- *Ash percent.* Shows the activated carbon's residue.
- *Iodine number.* An important parameter to be determined because it can indicate a carbon's ability to adsorb low molecular weights and be regenerated.
- *Molasses number, value, and decolorizing index.* For indication of a carbon to adsorb high molecular weights.
- *Pore size.* Used to obtain a carbon which can adsorb specific molecules.

The relationship between a carbon's adsorbability of a substance and that substance's concentration in a waste stream (or other liquid) is the adsorption isotherm. Adsorption isotherms are determined at constant temperatures and controlled operating conditions (pH, flow rate, and so on). From the adsorption isotherm, the amount of carbon required to effect the desired effluent characteristics can be estimated.

After estimates have been determined in the laboratory, pilot plant studies should be conducted to determine:

- carbon type, size, dosage
- bed dimensions
- effluent characteristics
- hydraulic characteristics
- dosage requirements
- contact time
- pretreatment requirements
- other effects

These other effects may include bacterial growth on the carbon bed, filterability, hydrogen sulfide generation, and pH and temperature effects on adsorption. Figures 2.13 and 2.14 are examples of two configurations of granular-bed carbon pilot plants.

Breakthrough curves, a plot of wastewater constituent concentration versus wastewaters treated, are used in determining the durability of a carbon with respect to the operating conditions it will encounter at a full-scale treatment facility.

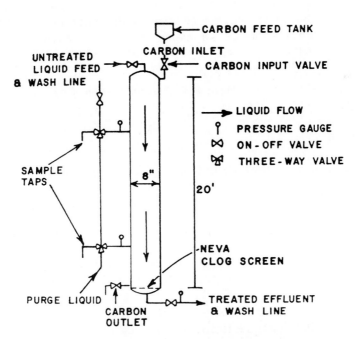

Figure 2.13 Fixed-bed carbon pilot plant.

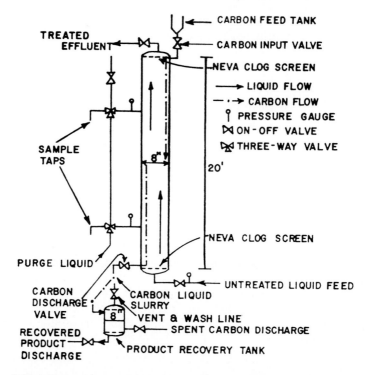

ONE FOOT SECTION OF CARBON PERIODICALLY REMOVED FROM
BOTTOM WHILE AN EQUAL AMOUNT OF FRESH CARBON IS
ADDED AT TOP, CARBON REMOVED IS WASHED TO RECOVER
PRODUCT.

Figure 2.14 Pulsed-bed carbon pilot plant.

Costs

Depending on the type and mode of carbon application, capital and operating costs for the system will vary. A batch operation, such as powdered carbon added to a waste stream as illustrated in Figure 2.15, would require a significantly smaller capital investment than the construction of carbon-bed adsorbers. However, carbon applied in this manner is not recoverable. Thus, the total carbon cost over the life of the plant may prove to be uneconomical.

Carbon systems' costs, like most treatment systems, are determined based on a unit cost per unit flow rate. Usually, then, as flow rates decline, unit costs increase per unit flow. The relative costs of several carbon adsorber systems are given in Table 2.4. These costs include construction, equipment, and carbon for equal treatment volumes.

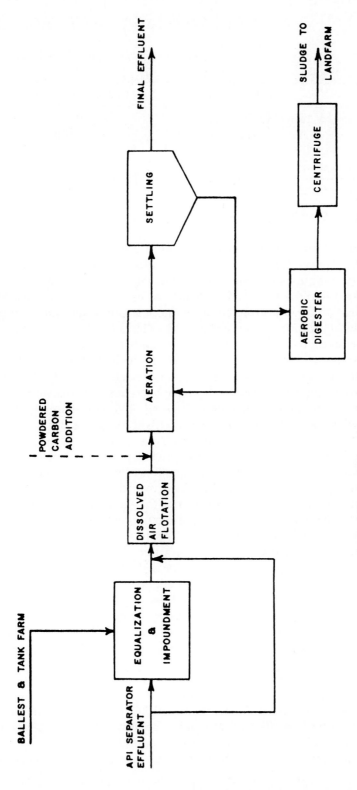

Figure 2.15 Schematic of a refinery wastewater treatment plant using powdered activated carbon.

TABLE 2.4   RELATIVE COSTS OF SEVERAL CARBON ADSORBER SYSTEMS

System	Cost Adjustment Factor
Upflow Countercurrent Packed Bed	1.0
Upflow Countercurrent Expanded Bed	1.0
Downflow Parallel	1.0
Downflow Series	1.28

RAW WATER/SEWAGE APPLICATIONS

As a result of an industrial spill or cleanup operation, an intermittent and intense taste and odor problem may be created. Persistent problems may be caused by continuous discharge of process wastes. Phenols and related compounds are often the source of medicinal tastes—tastes which are intensified by chlorination. Hydrocarbons from refinery wastes often can be recognized by an oily film on the water surface. As little as 0.025 to 0.050 mg per liter of such wastes have caused taste and odor outbreaks. Some chemical wastes, such as ethyl acrylate and N-butyl mercaptan, have been detected by the threshold odor test at 0.006 and 0.007 mg per liter, respectively. Odors resulting from pulp and paper wastes have been described as sulfite or paper-mill odors. Even though such terms are neither very descriptive nor precise, anyone who has been inside a paper mill will recognize the typical odor of such wastes. Wastes from the metal-processing industry may cause tastes, but not odors. Zinc, copper, and other metals produce characteristic tastes.

Sewage contains a mixture of organic materials. In sewage treatment, some of these compounds may be partially oxidized and produce odor. When sewage is chlorinated to control bacteria, the effluent may have a chlorine-type odor, due to the formation of chlororganic compounds. Sewage polluted water contains a relatively high concentration of nitrogen compounds. These may release ammonia which reacts with free chlorine to produce nitrogen trichloride. This compound has a very persistent and irritating odor.

Miscellaneous sources of taste and odor may include pest control, where various chemicals have been employed, such as insecticides, fish poisons, and herbicides. Careless application has permitted some of these pesticides to enter streams which serve as sources of public water supplies. Usually the taste and odor problems have been associated with the solvents used to disperse the chemicals.

The use of activated carbon, a material specially treated to produce a surface condition of great adsorption capacity, is very effective in removing most tastes and odors. It is generally used in the form of a fine powder in dosages ranging from 10–50 lbs/thousand gallons. At times, as much as 300 lbs/thousand gallons are required for brief periods.

Application of carbon to the raw water will minimize the decomposition of sludge deposits in sedimentation basins. However, carbon seems to be more effective in removing taste and odor when it is on the surface of the filters where intimate contact with the filtering water is assured. Because of this, the practice has developed of adding small, uniform doses of carbon to the raw water, and adding varying and relatively larger doses to the settled water.

Many wastewater plant operators prefer to apply relatively large amounts for brief periods of time to individual filters immediately after each backwashing. This procedure has the advantage that an adequate, but not too great, amount of carbon is present on the filter bed throughout its operation period. Disadvantages are the difficulty of getting a uniform coating over the surface of the bed and the possibility of a decrease in the efficiency of taste and odor removal toward the end of the filter run.

Granular activated carbon is sometimes used as a filter, with the water passing through the beds at the rate of 2–4 gpm/sq ft. Since large amounts of activated carbon are quite effective in removing chlorine from water, the use of carbon filters is limited to applications before chlorination or when dechlorination as well as taste and odor removal are desired. The carbon granules become coated and the minute pores become clogged unless the water is clear. Therefore, carbon filters usually follow sand filters. Carbon filters must be washed at intervals but eventually their adsorptive power is exhausted and the carbon must be replaced or reactivated.

ADSORPTION WITH SILICA GEL

Silica gel is a hard, clear, and glassy substance with a chemical composition of approximately 100 percent. SiO_2 is very carefully prepared in order to give a definite physical structure. It is inert chemically and strong mechanically, so that large masses of the granular material may be employed without serious loss by attrition and may be continually heated without loss to the adsorptive properties. The extremely large surface of the pores with the addition of their capillary action in combination with its other properties make silica gel particularly suitable for a wide variety of purposes.

Industrial applications of silica gel may be grouped under three headings:

- Adsorption from the gas phase.
- Adsorption from the liquid phase.
- As a catalyst or as a support for catalytic agents.

Silica gel selectively adsorbs vapors of volatile liquids, such as water, acetone, benzene, petrol, and so on, from air and similar gases. If an air-vapor mixture of this type is passed through a bed of silica gel, vapor will be adsorbed until the break point when the air leaving will contain gradually increasing quantities of the vapor. The curve shown in Figure 2.16 shows the adsorption of water vapor from air under definite conditions of temperature, humidity, and rate of flow and indicates a number of points of general interest in connection with vapor adsorption. It will be noted that under these particular conditions, the break point is reached when the silica gel has adsorbed water from the air to the extent of 20 percent of its weight. On continuing to pass air through the silica gel, water vapor will be present to an increasing extent in the air as it leaves, and the two curves shown represent alternative methods of expressing the adsorption efficiency. The average efficiency is calculated from the ratio of the total amount of vapor adsorbed and the total amount of vapor present in the air-vapor mixture that has passed through the gel. The incremental efficiency, on the other hand, represents the ratio of the total amount of vapor adsorbed and the total amount of vapor present in the air-vapor mixture passing through the gel at any given moment. The incremental efficiency is the actual efficiency of adsorption measured at one particular instant during an adsorption cycle, while the average efficiency is the average for the whole period of the adsorption cycle. There is an applicable difference in the adsorptive capacity between these two curves and it is important to bear this in mind when considering efficiencies. High adsorption

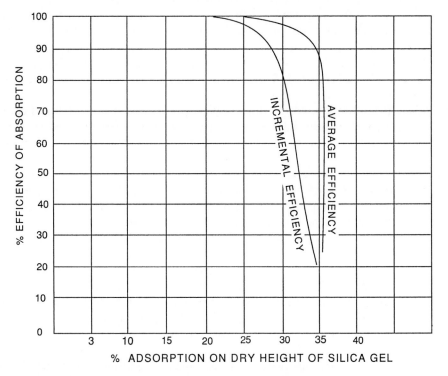

Figure 2.16   Adsorption of water vapor from air by silica gel.

figures may be obtained under certain conditions, based on adsorption for complete equilibrium under static conditions, but these may be misleading and do not represent results that are of value from the industrial point of view.

A number of arrangements can be adopted employing silica gel for adsorption purposes which will depend on the specific requirements such as the quantities, concentration, temperature, and so on for any given problem. A typical plant would consist of two adsorbers, each containing the requisite quantity of silica gel, arranged so that one is adsorbing while the other is being regenerated. Regeneration will comprise removal of the adsorbed material in such a way that the silica gel has its original low condensed vapor content and is ready for another adsorption cycle. The adsorbed liquid after removal from the gel may be condensed and recovered. Any number of adsorbers may be arranged so that they are worked in a convenient cycle of operations.

Adsorptive properties of silica gel from the gas phase afford a number of applications, such as drying, the drying and purifying of compressed gases, dehydration of coal gas, and so on. The treatment of compressed gases, such as compressed air, oxygen, nitrous oxide, carbon dioxide, and others, not only deals with the small residual amount of moisture remaining after compression, but also purifies the gas by removing contamination due to noxious vapors.

A special case of air drying by silica gel is the dehydration of process and instrument gases. When gas or air is passed over silica gel, the latter adsorbs water vapor and benzene. After a time, however, owing to its selective adsorptive properties, a certain amount of the adsorbed benzene may be replaced by water so that by varying the conditions it is possible that varying ratios of water and benzene adsorbed by the gel may be obtained in such applications. Results obtained from a plant operation showed that when obtaining the required dehydration of some 60 percent, the ratio of water to benzene adsorbed was approximately 3:1, while at the same time, naphthalene was almost completely removed from the gas, and the very slight reduction in the properties of the dehydrated gas was more than offset by the value of the benzene recovered. Apart from the advantages of dehydrating the gas to such an extent that water deposition in mains and services is prevented, the removal of napthalene is an advantage.

An example of the use of silica gel as a support for catalytic agents is platinized silica gel, or gel impregnated with another suitable catalyst. This is of great value in the contact process for the manufacture of sulphuric acid. The impregnated silica gel may be used to give rise to increased yields.

Table 2.5 gives a comparison of the treatment processes that separate organics from waste streams and can be compared to a carbon adsorption. Table 2.6 illustrates typical organics and their properties that may be collected on activated carbon.

TABLE 2.5 COMPARISON OF TREATMENT PROCESSES THAT SEPARATE ORGANICS FROM LIQUID WASTE STREAMS

Treatment process	Required feed stream properties	Characteristics of output stream(s)
Carbon Adsorption	Aqueous solutions; concentrations <1%; SS <50 ppm	Adsorbate on carbon; usually regenerated thermally or chemically
Resin Adsorption	Aqueous solutions; concentrations <8% SS <50 ppm; no oxidants	Adsorbate on resin; always chemically regenerated
Ultrafiltration	Solution or colloidal suspension of high molecular weight organics	One concentrated in high molecular weight organics; one containing dissolved ions
Air Stripping	Solution continuing ammonia; high pH	Ammonia vapor in air
Steam Stripping	Aqueous solutions of volatile organics	Concentrated aqueous streams with volatile organics and dilute stream with residuals
Solvent Extraction	Aqueous or nonaqueous solutions; concentration <10%	Concentrated solution of organics in extraction solvent
Distillation	Aqueous or nonaqueous solutions; high organic concentrations	Recovered solvent; still bottom liquids, sludges, and tars
Steam Distillation	Volatile organics, nonreactive with water or steam	Recovered volatiles plus condensed steam with traces of volatiles

TABLE 2.6 VARIOUS ORGANICS THAT MAY BE COLLECTED ON ACTIVATED CARBON AND RETENTIVITY

Compound	Formula	Molecular weight	Boiling point 760 mm C	"C" critical temperature	Approximate retentivity in % at 20C 760 mm
Methane Series	$CnH2n+2$				
Methane	$CH4$	16.04	−184	−86.7	1
Ethane	$C2H6$	30.07	−86	32.1	1
Propane	$C3H8$	44.09	−42	95.6	5
Butane	$C4H10$	58.12	1	153.0	8
Pentane	$C5H12$	72.15	37	197.2	12
Hexane	$C6H14$	86.17	69	234.8	16
Heptane	$C7H16$	100.20	98.4	266.8	23
Octane	$C8H18$	114.23	125.5	296.0	25
Nonane	$C9H20$	128.25	150.0		25
Decane	$C10H22$	142.28	231.0		25

TABLE 2.6 VARIOUS ORGANICS THAT MAY BE COLLECTED ON ACTIVATED CARBON AND RETENTIVITY (*continued*)

Compound	Formula	Molecular weight	Boiling point 760 mm C	"C" critical temperature	Approximate retentivity in % at 20C 760 mm
Acetylene Series	CnH2n-2				
Acetylene	C2H2	26.04	−88.5	36.0	2
Propyne	C3H4	40.06	−23.0		5
Butyne	C4H6	54.09	27.0		8
Pentyne	C5H8	68.11	56.0		12
Hexyne	C6H10	82.14	71.5		16
Ethylene Series	C2H2n				
Ethylene	C2H4	28.05	−103.9	9.7	3
Propylene	C3H6	42.08	−17.0	92.3	5
Butylene	C4H8	56.10	−5.0		8
Pentylene	C5H10	70.13	40.0		12
Hexylene	C6H12	84.16	64.0		
Heptlene	C7H14	98.18	94.9		25
Octalene	C8H16	112.21	123.0		25
Benzene Series	CnH2n-6				
Benzene	C6H6	78.11	80.1	288.5	24
Toluene	C7H8	92.13	110.8	320.6	29
Xylene	C8H10	106.16	144.0		34
Isoprene	C5H8	68.11	34.0		15
Turpentine	C10H16	136.23	180.0		32
Naphthalene	C10H8	128.16	217.9		30
Phenol	C6H5OH	94.11	182.0	419.0	30
Methyl Alcohol	CH3OH	32.04	64.7	240.0	15
Ethyl Alcohol	C2H5OH	46.07	78.5	243.1	21
Propyl Alcohol	C3H7OH	60.09	97.19	263.7	26
Butyl Alcohol	C4H9OH	74.12	117.71	287.0	30
Amyl Alcohol	C5H11OH	88.15	138.0	307	35
Cresol	C7H7OH	108.13	202.5	122	30+
Menthol	C10H19OH	156.26	215		20
Formaldehyde	H.CHO	30.03	−21.0		3
Actaldehyde	CH3CHO	44.05	21.0	188	7

SOLVENT RECOVERY

Basic stages to consider for a carbon adsorption solvent recovery plant include:

- Efficient collection of solvents evaporated in the process.
- Adsorption process during which the carbon is charged with solvent.

- The steaming or stripping process, and collection of stripped solvents (recovery).
- The drying and cooling of the carbon in preparation for the next adsorption charge.

In any recovery system the most important factor is to ensure that the maximum solvent is collected efficiently and safely from all sources of evaporation. Carbon adsorption process recoveries of over 99 percent of the solvent in the air stream passing to the carbon adsorption plant are readily achieved in present-day equipment. Experience shows that the greater loss of solvent occurs in the handling and manufacturing process, rather than in the recovery plant. Slight losses of solvent are unavoidable no matter how many precautions are taken to prevent incidental losses. In theory, the problem of solvent recovery is simple. If process could be carried out continuously in gas-tight equipment, it would be possible to recover the solvents efficiently and cheaply by direct condensation. Direct condensation is employed in many processes including distillation, extraction, and so on, but in a large number of processes it is impossible to work continuously in a completely closed system. Even in a simple tank, every filling operation is attended by the venting of air saturated with vapor at the temperature of its contents during the displacement of the tank contents.

Some industrial processes are of such a nature that it is impracticable to work with an entirely closed apparatus, and continuity can only be secured over short periods. Processes such as coating, impregnating, film casting, and printing are such that an ingress of air is almost unavoidable, and, indeed, is often necessary to ensure that the rate of evaporation or drying can be controlled or effected.

For reasons of safety, when using flammable or toxic solvents such as acetone, benzene, carbon disulphide, carbon tetrachloride, toluene, and so on, adequate ventilation must be used to prevent solvent concentrations from reaching dangerous proportions inside machinery and workrooms where the solvent is evaporated. (See Chapter 7 for safety.)

The approximate lower and upper explosion limits in air for some commonly used solvents are given in Table 2.7.

The quantity of air used to entrain the solvents being evaporated from any particular source of equipment will depend on the manufacturing process, but as a general rule the volume of air must be sufficient to assure that the solvent concentration will be below the lower explosive limit. Concentrations in the order of 25 percent to 50 percent of the lower explosive limit are used in industrial practice.

Size and cost of a carbon adsorption plant is directly dependent on the volume of air to be handled. Therefore, it is important to build up the solvent concentration to as high a level as can safely be tolerated. The solvent concentration may be increased in the equipment of many processes by the use of air recirculation. This is often advantageous for increasing the rate of evap-

TABLE 2.7 EXPLOSION LIMITS OF SOME FLAMMABLE COMPOUNDS IN AIR

	Lower explosive limit (percent by volume in air)	Upper explosive limit (percent by volume in air)
Benzene	1–4	7–1
Toluene	1–4	6–7
Xylene commercial	1	6
ortho		
meta	1	7
para		
n-pentane	1–4	8
n-hexane	1–2	6–9
n-heptane	1–2	6–7
Carbon disulphide	1–25	44
Ether	1–85	48
Methyl ethyl ketone	1–8	10
Acetone	3–0	11
Methanol	7–3	36
Ethyl alcohol	4–3	19

oration, since air may be blown directly onto the surface of the products from which the solvent is evaporating, thus increasing the rate of evaporation and permitting a higher production rate for a given size.

The use of a high air-recirculation rate is advantageous in many processes, including printing and textiles proofing. But in many industries the use of highly turbulent air for drying would spoil the product. This is especially true in the film casting industry where rapid evaporation would cause the formation of bubbles and rippling of the film surface. Also with some processes higher rates will cause more solvent to be evaporated unnecessarily, and care must be taken to avoid this.

The range of processes from which solvents are evaporated is extensive, and each process presents a specific problem for the designer of the recovery system. As long as the air volume employed for suction is sufficient to assure that the lower explosive limit is not reached, flame propagation or explosion need not be a danger. Evaporation of solvents from industrial plants, rubber-spreading machines, mixers, dry-cleaning plants, printing, and so on is not uniform. Equipment is started and stopped throughout the operation. Consequently, it is important not only to provide a sufficiently large air volume, but to take into account the fluctuations and provide additional safety precautions. Fluctuations which take place in the rate of solvent evaporation may be as much as eight or ten times the average evaporation rates. In such circumstances it would be uneconomical to build a plant ten times the average size and capacity to meet this condition.

A safe method of assuring controlled suction in the right place is to provide a hood over every machine where solvent is evaporated, a valve to shut off the suction when the machine is not in use, and a flow meter and valve to regulate the volume of air drawn off to suit the rate of evaporation. In this way a proper distribution of the air volume over many machines in operation can be maintained. Many larger installations are outdoors and do not require this.

Additional safeguards can be made by dividing up the suction system between the equipment by using flame arrestors and pressure release valves on each hood, flame arrestors at the junction of main suction lines, and in certain cases rupture disks facing into the open air so that any pressure may be released safely.

Flame arrestors are based on the rapid removal of the heat of combustion during the passage of the burning gases so that they are extinguished. Suitable designs are constructed of sheet metal curved to produce a large number of air passages. A water-cooled heat exchanger of the extended or finned-tube variety is a first-class flame arrestor and is often installed on recovery systems to cool the solvent-laden air prior to the carbon adsorption plant.

Reducing the volume of hoods over the solvent-evaporation apparatus will also increase the safety factor. The smaller hood will give a higher air velocity for a given air rate and prevent building up of pockets of solvent as well as reducing the solvent capacity of the hood. If for any reason an explosive mixture was built up in a small hood, the total quantity of vapor which could be ignited would be small and would not be a source of danger, especially if equipment were isolated from the rest of the plant by means of flame arrestors and explosion or rupture disks.

The design and distribution of suction lines for solvent-laden air is an important part of the recovery process, and as much care must be given to this part of the installation as to the production equipment and carbon adsorption recovery plant. The installation of a well-designed duct system provides a major contribution to the efficiency of solvent recovery as well as the assurance that the plant will be run under the safest controlled operating conditions.

3

Treatment of Liquids

Carbon treatment has been used in numerous positions in the sequence of plant operations. Sometimes, it is used early in a process to remove gross quantities of a contaminant or it has been used as a final step for improving product quality by removing trace components and as a polishing operation, or in wastewater treatment as a tertiary or advanced treatment process.

Both powered and granulated activated carbon have been used successfully and there are advantages and disadvantages to using each type. However, there has been a trend to convert existing powdered carbon operations to granular systems and to use granular carbon in new systems. Reasons for this are:

- Decreasing price differential between the two carbons.
- Problems with disposal of powdered carbon and filter cake.
- Lack of a competitive thermal reactivating system for powdered carbon. Currently, work is being done to develop reactivating equipment for powdered carbon.
- Greater labor requirement for the powdered carbon system.
- Higher product losses per weight of carbon used.
- Inefficient use of the carbon.

Although there are many similarities in the carbon systems for the different services, there are also many differences. The following are some points of comparison between the two applications:

	Application Water/Waste	Chemical/Food
Typical Range of Superficial Contact Times, minutes	5–100	30–400
Typical Linear Flow Rates, m/hr/gpm/ft²	10–20 4–8	1–5 0.4–2.0
Flow Rates Encountered, m³/hr gpm	2.27–110,000 10–350,000	0.23–114 1–500
Carbon Bed Depth, m	0.5–10	3–20
Backwashing and Air Scouring	Yes	No
Viscosity Range, cp	0.5–2.0	1.0–40.0
Biological Activity in Carbon	Sometimes	Rarely
Temperature Range, °C	25–45	45–100
Range of Removal of Impurity, %	30–99.99	30–98
Carbon Dosage, lb/1,000 gal kg/m³	1–300 0.012–3.6	50–400 0.64.8

On a theoretical basis, it is not possible to predict how effective carbon will be in treating a given liquid, or the conditions under which the carbon will do the most effective job. Therefore, it is usually necessary to conduct laboratory tests to determine:

- To what extent removal of the component(s) or adsorbate is possible.
- Carbon dosage, or g carbon/g of liquid.
- Amount of carbon required to be onstream to efficiently remove the adsorbate.
- Effect of linear flow rate and superficial contact time on performance of the carbon.
- Type of system to install, that is, fixed beds or pulse beds—singly, in series, or in parallel.
- Effect of temperature and/or pH on the adsorption capacity.

A complete laboratory investigation would generally consist of two parts: preliminary isotherm tests would be performed to demonstrate the feasibility of granular carbon treatment and laboratory column tests would be conducted to obtain data to be used in designing the full-scale plant. Unlike column tests in water-related applications, which may take several months, most column runs for chemical applications can be completed in less than a month.

ADSORPTION ISOTHERMS

An adsorption isotherm is a simple method of determining the feasibility of using granular activated carbon for a particular application. A liquid-phase isotherm shows the distribution of adsorbate (that which is adsorbed) between the adsorbed phase and the solution phase at equilibrium. It is a plot of the amount of adsorbate adsorbed per unit weight of carbon versus the concentration of the adsorbate remaining in solution.

Straight-line plots can generally be obtained by making use of the empirical Freundlich equation, which relates the amount of adsorbate in the solution phase to that in the adsorbed phase by the expression:

$$x/m = kc^{1/n} \tag{3-1}$$

where

x = amount of adsorbate adsorbed
m = weight of carbon
x/m = concentration in the adsorbed phase, that is, the amount of adsorbate adsorbed per unit weight of carbon
c = equilibrium concentration of adsorbate in solution after adsorption

k and n are constants. Taking the logarithm of both sides we obtain:

$$\log x/m = \log k + 1/n \log c \tag{3-2}$$

This equation is a straight line whose slope is $1/n$ and whose intercept is k at $c = 1$. Therefore, if x/m is plotted against c on log-log paper, a straight line will normally be obtained. However, there are occasions, as explained later, when this is not true. The straight and the curved isotherm lines provide valuable information for predicting adsorption operations.

EXPERIMENTAL CONDITIONS AND VARIABLES

Particle Size of Carbon

In liquid-phase applications, transfer of the adsorbate from the bulk solution to the carbon particle must proceed through two stages:

- Transfer of the adsorbate from the bulk liquid to the surface of the carbon particle.
- Migration of the adsorbate from the surface of the carbon to the adsorption site within the particle.

Kinetic experiments demonstrate that the transfer of adsorbate is appreciably more rapid than the migration of the absorbate; so the latter is normally the primary rate-determining step. The rate of adsorption will vary with the diameter of the carbon particles used in testing.

To increase rates of adsorption and decrease the time necessary to complete the isotherm, it is recommended that the granular carbon be pulverized so that 95 wt % will pass through a 325-mesh screen. Such pulverization does not significantly increase the surface area. The increase in the surface area, in most cases, is less than 1 percent, as the vast majority of surface area is contributed by the pore walls rather than by the external surface of the carbon particles.

Liquid Temperature

Adsorption efficiency is usually a function of system temperature. For laboratory evaluation of granular activated carbon of liquid-phase applications, it is recommended that the temperature of the existing process stream be determined first. If the desired degree of adsorption occurs at this temperature, then the plant operating parameters will not have to be altered. If, on the other hand, the desired degree of adsorption does not take place at the existing process temperature, then a higher or lower temperature should be evaluated. The selection of the temperature will depend on:

- viscosity
- thermal stability characteristics of the test liquid
- feasibility of changing the process temperature in the plant

Liquid pH

Adsorption capacity can also be a function of the pH liquid. Normally, when performing laboratory evaluations, the pH of the process stream is used. If the desired degree of adsorption does not take place at the process pH, then various pH levels should be investigated. Care must be taken when adjusting the pH of a process stream to make sure that the change in pH does not degrade or decompose a particular product or material. When treating liquids with a low pH, it is advisable to use an acid-washed carbon, since a portion of the ash constituents of nonacid-washed activated carbon will be solubilized under acidic conditions.

Contact Time

Contact time is critical to the adsorption process. It should be sufficiently long to allow an approach to adsorption equilibrium. A preliminary experiment can be performed to determine the contact time required to attain equilibrium.

Normally, 0.5-g portions of pulverized activated carbon should be added to several 100-g portions of the test liquid (when the density of the liquid is near that of water, 100 ml of the liquid, carefully measured, may be used) and each portion agitated for various time periods at the process temperature and pH. For example, portion 1 is agitated for 30 minutes portion 2 for one hour, portion 3 for two hours, and so forth. At the end of each contact time, the carbon is removed by a suitable means such as filtration or centrifugation. If the quantity of adsorbate remaining in solution is plotted as a function of time, a curve similar to Figure 3.1 should be obtained. It is apparent from Figure 3.1 that the contact time required to reach equilibrium for this particular liquid is about three hours. Isotherm experiments for estimating the required carbon dosage then must be performed with a minimum of three hours contact time.

Carbon Dosage

To obtain a meaningful isotherm, as wide a range of carbon dosages as practical should be used. Recommended dosages are 0.05, 0.1, 0.2, 0.5, 1.0, 2.5, 5.0, and 10.0 g of carbon per 100 g of the test liquid.

Separation of Carbon from Liquid

Prior to analysis of the treated liquid, the carbon must be removed from the liquid. Carbon can be removed by filtration through a 0.45-μ pore-size filter pad. In some cases, a smaller pore-size filter pad may be necessary to remove all carbon particles, and the filter pad which is used must be compatible with the treated liquid. The filtration rate may be increased by using heat or pressure. If the liquid is so viscous that filtration is difficult centrifugation can sometimes be used as a suitable separation.

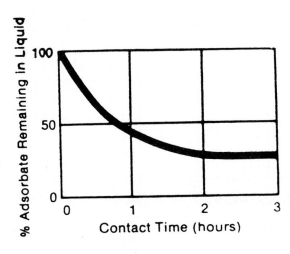

Figure 3.1   Contact time.

Adsorbate Removal

After the carbon has been removed from the treated liquid, the amount of adsorbate remaining in solution can be determined. Any analytical method or instrument that can determine the quantity of the particular adsorbate remaining in the test liquid may be used. Techniques include titration, spectrophotometry, gas/liquid chromotography, and total organic carbon analysis.

Isotherm Significance and Adsorption Capacity

The isotherm will determine whether the desired degree of adsorbate removal is possible with the particular activated carbon tested. If a vertical line is drawn from the point on the horizontal axis (Figure 3.2) corresponding to the influent

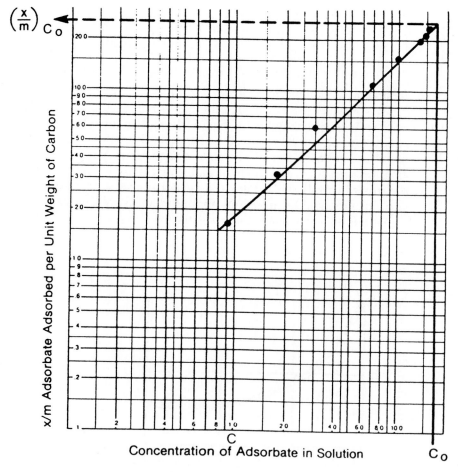

Figure 3.2 Straight-line isotherm plot.

concentration (c_o) line, and the best line through the data is extrapolated to intersect this (c_o) line, the (x/m) value at the point of intersection can be read from the vertical scale. This (x/m) value represents the amount of adsorbate adsorbed per unit weight of carbon when that carbon is in equilibrium with the influent concentration. This is the ultimate capacity of the carbon at these conditions. The ultimate capacity may or may not be realized in column operation, since the adsorption rate is limited. Thus, the superficial contact time in a column operation is important.

From the value of (x/m), the quantity of liquid treated can be calculated using the following formula:

$$W_{c_o} = \frac{(x/m)_{c_o}(W)}{c_o} \qquad (3\text{-}3)$$

where

W_{c_o} = theoretical weight of liquid treated per gram (or unit weight) or carbon

$(x/m)_{c_o}$ = capacity per gram (or unit weight) of carbon at the influent concentration

W = weight of liquid used in the isotherm test

c_o = influent concentration

If the liquid was measured by volume, the same calculation should be used by substituting V_{c_o} and V in the proper places.

$$V_{c_o} = \frac{(x/m)_{c_o}(V)}{c_o} \qquad (3\text{-}4)$$

where

V_{c_o} = theoretical volume of liquid treated per gram (or unit weight) of carbon

$(x/m)_{c_o}$ = same as Equation 3-1

V = volume of liquid used in the isotherm test

c_o = same as Equation 3-1

The performance of two or more carbons in the same application can be compared by examining their isotherms. Usually, granular carbon with the higher $(x/m)_{c_o}$ value would be preferred for an application.

Nonlinear Isotherms

Ideally, straight-line isotherm plots are obtained. However there may occasionally be departures from linearity. A curve as shown in Figure 3.3A may be obtained if a nonadsorbable impurity is present in the liquid being treated. For such situations subtracting c_1 from c_o and replotting the isotherm will usually yield a straight line.

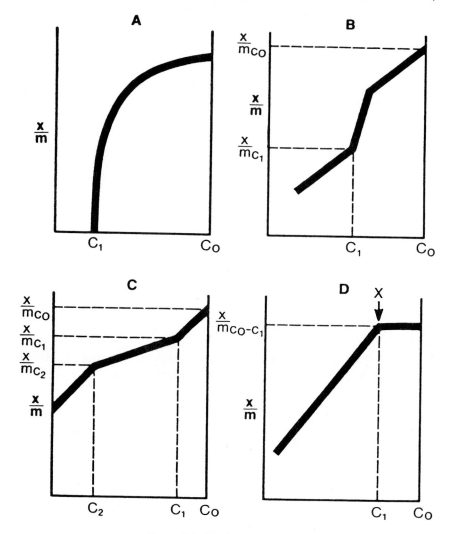

Figure 3.3   Nonlinear isotherms.

A sudden change in slope (as in Figure 3.3B) indicates two components (adsorbates) are present which are adsorbed at different rates with markedly different capacities. Figure 3.3C, similar to that in Figure 3.3B, illustrates a situation in which three compounds (adsorbates) are present, and all are adsorbed in different degrees. When many adsorbate species are present, but all are adsorbed equally, straight-line isotherms still will be obtained. Figure 3.3D indicates that the adsorbate has reached its maximum surface coverage at capacity X. Therefore, any increase in concentration of adsorbate does not result in increased capacity.

From an isotherm test, it is possible to determine whether a required adsorbate removal can be accomplished and obtain the ultimate capacity of the granular carbon for that application. From the capacity figure, an estimate of the minimum granular carbon usage rate necessary to meet the treatment objective can usually be obtained. Isotherm tests also afford a convenient method for comparing different carbons and for investigating the effects of pH and temperature. The lowest possible carbon usage rate is predicted by a properly run isotherm test. If a lower usage is observed in column tests or in the plant, then one or more of the following is true:

- Isotherm test was not at equilibrium.
- Interpretation of the isotherm was done incorrectly.
- An error in the analytical techniques.
- Carbon fines may not have been completely removed from the solution.
- Liquids tested in the isotherm test and in the column tests were not identical.
- A bacteria strain present in the column which removes adsorbate. This phenomenon is usually limited to water applications.

ADSORPTION IN COLUMNS

If isotherm studies indicate that the liquid can be treated to the desired purity level at a reasonable dosage, then the next step is to evaluate the liquid in a dynamic test.

Adsorption on granular activated carbon is a diffusion process consisting of the following steps:

- Bulk diffusion of the adsorbate from the liquid to the film around the carbon particle.
- Diffusion through the film.
- Internal pore diffusion to the adsorption sites.

In multicomponent systems there is the added step of competition for site results in displacement of previously adsorbed, but less strongly held, components.

Ideally, it would be desirable to be able to mathematically model the carbon performance from the equilibrium data. In studies used to develop models to date, one or more of the following elements is usually present in the study or assumed in the model:

- Uniform carbon particle size.
- Ideal isotherm.
- Dilute solutions.

- Single component.
- Adsorbate molecule was too large to enter any pore, so there was no pore diffusion, thus film diffusion was controlling.
- Flow rate was high so pore diffusion was controlling.
- Comparisons of predicted versus actual data were made on the early part of the breakthrough curve before the carbon was saturated.
- Systems with relatively short mass-transfer zones (MTZ) were studied.

There is no model presently available that can be used universally. Until such time as one is developed, it will be necessary to test each liquid in a dynamic system and use empirical methods to design the plant-scale units. Prior to conducting a column test, considerations consist of:

- Location of test, that is, in the plant or in the laboratory.
- Size and type of pilot system, that is, column diameter, quantity of carbon, and fixed or pulse bed.
- Carbon type and particle size.
- Linear velocity of liquid in the carbon bed.
- Temperature.
- pH.

Testing

Generally, the tests can be controlled better in the laboratory. However, tests should be conducted in the plant if any of the following conditions exist:

- The liquid to be treated would deteriorate or change in characteristics during shipment and/or storage.
- Carbon dosage is so low that large quantities of liquid would be required to conduct the test.
- Properties of the liquid stream being treated vary widely. This is usually not nearly as important a factor as it is in wastewater applications.

Generally, experience may serve as a guide to selecting the size of the system. If there is no previous history for the liquid, it is best to overdesign the capacity. In the laboratory, the columns are usually 2.5–5.0 cm in diameter by 1–1.5 meters deep. In the plants, the size is usually 5–30 cm in diameter. If the columns are fixed beds, they range from 1–2 meters high and are connected in series. If a pulse bed is used, the bed depth is usually 10–15 meters. Assuming there are sufficient sample points in both types of systems, the same quality of data can be obtained. The pulse bed has the advantage of easy carbon addition and removal.

Carbon Type and Particle Size

Carbon type to be used will usually have been determined during the iso-therm testing. Since some carbons used in chemical operations are used on a throwaway basis or can be reactivated to near virgin activity, testing with virgin carbon is representative. Because pore diffusion is a major contributor to the overall diffusion rate, selection of the particle size could be critical. The smaller the particle, the faster the diffusion, and thus, the shorter the mass-transfer zone. This is illustrated in Figure 3.4 and it is obvious that in a single fixed bed the column having the smaller-diameter carbon particles would treat a greater volume of liquid if it had to be taken offstream at c/c_o = 20 percent. However, if the full-scale carbon system is designed to remove only saturated carbon from the system by using beds in series or a pulse bed, the carbon dosage would be the same for either particle diameter carbon.

Velocity of Liquid in the Bed

Through the years, there have been many studies conducted and reported as to if and how linear velocity affects diffusion rate. There are data showing that at the same contact time, but different linear velocities, there is no difference in the performance of a carbon system.

It is obvious then that the effect of linear velocity on the diffusion through the film around the particle and the ratio of the magnitude of the film diffusion to the pore diffusion are the factors that determine the effects, if any, that occur. Therefore, the linear velocity cannot be ignored completely when evaluating a system.

Systems at the higher linear velocity (LV) treat more liquid per volume of carbon at low-concentration levels and the mass-transfer zone (MTZ) is

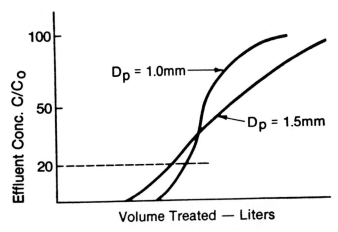

Figure 3.4 Effect of particle size on breakthrough curve.

shorter. The MTZ is that length of the adsorbent bed where the adsorbate concentration in the liquid varies from the influent concentration to zero or the specification value. At the same contact time, the single fixed-bed system at the higher LV stays onstream about 15 percent longer.

Temperature and Pretreatment

As adsorption is a function of diffusion rate and as diffusion is affected by the liquid viscosity, columns should be operated at the plant process or isotherm temperature to eliminate this variable.

If the isotherm investigation indicates that a pH adjustment is necessary before carbon treating, then the same pH adjustment should be made to the liquid before granular carbon column studies. In cases where suspended matter is present in the test liquid, it should be removed by filtration before the liquid is pumped to the columns. If the test liquid has been stored at a low temperature for preservation purposes, subsequent heating to room temperature or higher may result in degassing the liquid. When these conditions exist, the liquid must be deaerated prior to pumping it through the carbon columns.

A common practice in adsorption column work which leads to poor adsorption results is that the carbon is not deaerated prior to the adsorption tests. If this is not done, air pockets form in the column and result in channeling, high-pressure drops, and premature breakthrough of adsorbate. The time required to deaerate carbon is a function of the liquid temperature and is approximately as follows:

Liquid Temperature	Time to Be 90% Wetted
25°C	2–3 days
100°C	2–3 hours
1,000°C	Instantaneous

Carbon should be prewetted prior to being placed in the test columns. Backwashing the carbon at low rates (2.5 m/hr) does not remove the air. Rates that would expand the bed 50 percent or 15–30 m/hr, are required. The liquid used for prewetting can either be water, if it is compatible with the liquid to be treated, or a batch of the liquid to be treated which has been purified previously.

Column Operation

The ideal column run consists of the following:

- The column setup (either fixed bed or pulse bed) would be operated until steady state was reached.
- Carbon columns would be rotated at a rate to simulate various dosages.
- Operation occurs at various linear velocities.

- Feed concentration or color is kept constant. When the data are analyzed, the feed concentration would then be varied as it would be varied in the plant.

Actual laboratory tests are usually conducted as follows:

- From 3–20 meters of carbon would be used, contained in two to twelve columns varying from 0.5–2 meters in depth.
- The columns vary in diameter from 2.54–30 cm in diameter.
- The flow is generally limited to the rate anticipated in the plant. Occasionally, two flow rates are tested.
- Virgin carbon is used.
- In most cases, all of the carbon is never completely saturated. In fact, most of the time, even the first bed of carbon is not saturated.
- The quality of the feed varies.

Column tests are intended to obtain breakthrough curves showing how the concentration of the effluent vanes varies with the volume of liquid treated. Some typical examples of breakthrough curves are shown in Figure

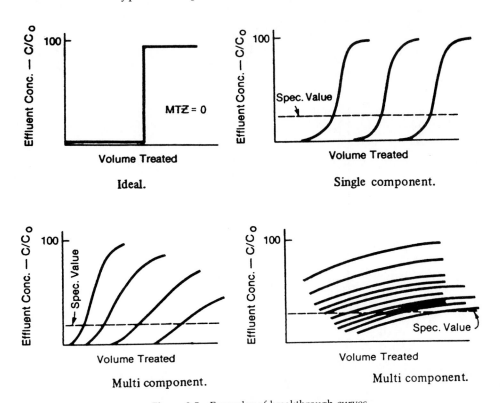

Figure 3.5 Examples of breakthrough curves.

3.5. Mathematical modeling of adsorption has not reached the point where it can be easily used as a design tool.

SYSTEM SELECTION

There are three types of carbon systems: (1) fixed beds, (2) pulse beds, and (3) fluidized beds, and these can be used singly, in parallel, or in combination. The majority of systems now onstream are either fixed or pulse beds.

The two basic types of adsorbers which can be designed to operate under pressure or at atmospheric pressure are the moving or pulse bed and the fixed bed. Either can be operated as packed or expanded beds.

In the pulse bed shown in Figure 3.6, the liquid enters the bottom cone and leaves through the top cone. The flow of liquid is stopped periodically, spent carbon is withdrawn (pulsed) from the bottom, and virgin or reactivated carbon is added into the top of the adsorber.

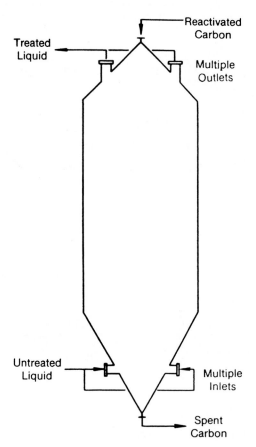

Figure 3.6 Moving- or pulse-bed adsorber.

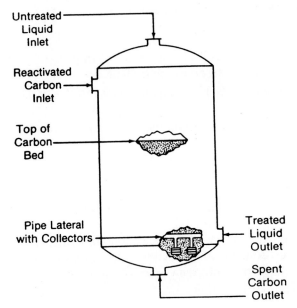

Figure 3.7 Fixed-bed adsorber (pressure).

In a fixed-bed adsorber shown in Figure 3.7, the liquid passes through the carbon until the carbon is spent or the effluent specification is reached. The entire volume of carbon is removed from the adsorber and virgin or reactivated carbon is placed in the adsorber. Some of the advantages and disadvantages of each system are:

FIXED-BED SYSTEM (Nonexpanded)

Advantages:

- Has lowest building profile.
- Requires least operator attention.
- Has fewer carbon fines in the effluent.
- Can be inspected internally each time the adsorber is emptied.
- Requires lower inlet pressure to pass through the system.

Disadvantages:

- Requires the greatest amount of plot area.
- Has higher capital investment than a pulse-bed system for the same amount of carbon onstream.
- In many cases, effluent quality can vary from no adsorbate in the effluent when a fresh adsorber is placed onstream to specification value when a spent adsorber is taken offstream. This is discussed later.

PULSE-BED SYSTEM (Nonexpanded)

Advantages:

- Requires the least area. This is probably the most important consideration when putting a system in an existing plant.
- Can be controlled to have the effluent quality close to the specification value.
- Has the lowest carbon dosage in applications having a long adsorption wave front.

Disadvantages:

- Cannot treat liquids containing significant quantities of suspended solids.
- Adsorber is not emptied in normal operations; therefore, inspection and/ or repairs can only be done by taking an adsorber off line and emptying the entire contents of carbon into temporary containers.
- Effluent contains carbon fines after each pulsing operation.

UP-FLOW EXPANDED BEDS

- The advantage claimed for this type of adsorber is that suspended solids in the liquid will pass through the bed. However, since there are different types of solids, a pilot test should be made to determine if the solids in the particular system will indeed pass through the bed of carbon. Also, special attention must be given to the underdrain design in a fixed-bed system because of some types of solids could plug it.

Either of two types of adsorbers can be arranged in various ways. These are shown in Figure 3.8 and are:

- Single adsorber.
- Two or more adsorbers in series (fixed bed only).
- Two or more adsorbers in parallel.
- Four or more adsorbers in parallel series.
  Selection of a particular system is generally based on the following:

SINGLE FIXED BED

- The carbon exhaustion rate is low and the cost of replacing or reactivating the carbon is a minor operating expense.
- The mass-transfer zone is short; that is, saturation of the carbon occurs shortly after breakthrough.
- The investment required for a multiple-column system cannot be justified by the lower carbon makeup rate.

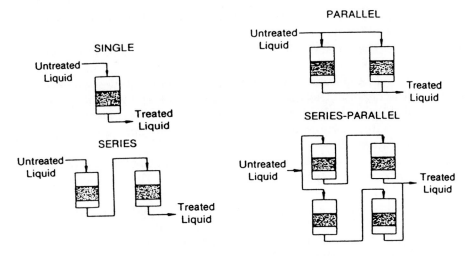

Figure 3.8 Fixed-bed system configurations.

FIXED BEDS IN SERIES

- The performance is such that the saturation of the bed occurs a long time after breakthrough; that is, the mass-transfer zone is long.
- It is economically attractive to have more stages to more completely exhaust the carbon.
- Influent concentration fluctuates widely.

FIXED BEDS IN PARALLEL

- The flow rate is high and the size of the adsorbers in a single pass would be too large to be economical or feasible.
- Space limitations prevent the use of large-diameter or extremely high columns.
- The adsorption process must be continuous. It is necessary to operate with a minimum pressure drop.
- Blending of product is desirable.

FIXED BEDS IN PARALLEL SERIES

- Flow rate is high.
- It is desirable to have a lower carbon dosage than possible with parallel only operation.

PULSE BEDS

- There is an incentive for approaching true countercurrent operation, especially when the mass-transfer zone is extremely long.

- The capital investment is usually lower than that required for fixed beds in series.
- There is a desire to minimize wide fluctuations in the effluent.

In applications in which a liquid is being purified with activated carbon, it is important to understand the variation in the effluent quality from the two types of adsorbers. In a fixed-bed system, the concentration of the effluent from the system will have the greatest fluctuation. A typical profile is shown in Figure 3.9.

In this case, an adsorber must be changed each time the effluent reaches the maximum specified level of impurity, c. Thus, the average quality of the product produced is significantly better than the specification. This results in quality giveaway. Two solutions to this would be to blend the product in tankage, or operate several columns in parallel and blend the effluents. Since a pulse-bed system is generally operated so that less than 10 percent of the carbon is pulsed at any one time, the variation in the effluent is much less and is shown in Figure 3.10. There is less product quality giveaway in this case. The magnitude of this variation can be changed by varying the size and/ or frequency of the pulse. In a continuous countercurrent system, the effluent quality would be constant as shown in Figure 3.11.

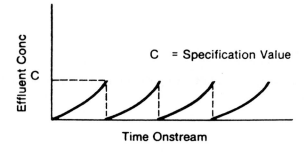

Figure 3.9 Effluent quality versus adsorber changeover.

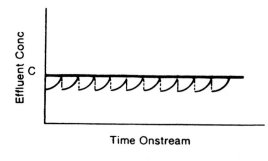

Figure 3.10 Effluent quality versus pulse frequency.

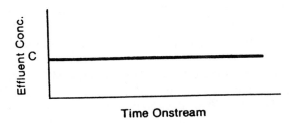

Figure 3.11 Effluent quality in a continuous countercurrent system.

SYSTEMS EQUIPMENT

Carbon Usage and Regeneration

Use of granular activated carbon for treatment of municipal and industrial wastewaters has developed in potable water applications and can be used to remove taste and odor as well as trace organics and chloroform. In the case of taste and odor removal, the carbon may last for a very long time before needing replacement. However, for trace organics or chloroform removal, carbon may need to be replaced more often depending on the concentration of contaminants. In either case, it is common to remove the sand from the existing gravity filters and replace it with 0.5—1.0 m (20–40 in.) of granular activated carbon. The resulting contact time is generally sufficient to achieve the desired removal efficiencies. The carbon can then serve as both a filter and an adsorber. This eliminates the need for a major capital expenditure for adsorbers and is simple and easy to operate.

Physical-chemical treatment can be used in place of biological treatment to obtain secondary treatment standards. Adding chemical coagulants to the wastewater to enhance solids settling and phosphate removal in the primary clarifiers is followed by adsorption and filtration of the clarified sewage through granular activated carbon. Some of the advantages include:

- Space requirements are generally smaller.
- Capital costs are lowered.
- Operating costs are lowered.
- No secondary sludge is generated.
- The ability to handle toxic spills and flow surges is present.

The carbon exhaustion rate for secondary applications ranges from 0.04–0.12 g/l (0.3–1.0 lb/1,000 gal). Design contact time will generally range from 15–30 minutes. Since the carbon is used as a filter as well as an adsorber, the adsorbers are operated downflow with an allowance for bed expansion during backwash. Both single-stage and two-stage designs have been used.

Granular activated carbon is also used for providing tertiary treatment following conventional biological treatment. In this case, the carbon exhaustion rate usually ranges from 0.01–0.06 g/l (0.1–0.5 lb/1,000 gal). Contact times will range from 15–30 minutes. As in the case of secondary treatment, the granular carbon can be used for filtration as well as adsorption, or prefilters can be provided.

Granular activated carbon has been used to remove organics from a wide range of industrial wastewaters including dyes, phenolics, benzene, and chlorinated hydrocarbons. Many of these organics are toxic and not biological process. The carbon exhaustion rate for industrial applications can range from as low as 50 kg/day (110 lb/day) to greater than 50,000 kg/day (110,000 lb/day) depending on the type and concentration of contaminant, effluent objective, adsorber configuration, and contact time. Contact times will generally range from 30–100 minutes, although longer contact times may be needed in special applications.

Moving beds, downflow fixed beds, and upflow expanded beds have all been used in industrial wastewater applications. Downflow fixed beds are the most common because of their inherent simplicity and reliability.

In most wastewater applications, the costs of virgin carbon usually prohibit its use on a throwaway basis for any but small installations. Thermal reactivation of hard, dense, coal-based carbons has proved to be economical and practical. Chemical regeneration is generally limited to applications where partial recovery of capacity is acceptable and regenerant disposal is not a problem. The most important factor affecting the costs of granular activated carbon treatment is the carbon exhaustion rate. The economics of using carbon on a throwaway basis can be compared with thermal reactivation on an off-site custom basis, an on-site basis, or a complete service basis.

The carbon exhaustion rate is expressed in terms of kilograms of carbon exhausted, pounds per day, or in grams of carbon exhausted per liter of water or wastewater treated (pounds per 1,000 gallons). This can be determined through laboratory or field pilot testing. The exhaustion rate is affected by the type and concentration of organics, flow rate, pH, contact time, and adsorber configuration. The design exhaustion rate is most important for selecting the optimum spent carbon-handling alternative.

In the throwaway carbon approach, virgin carbon is purchased in bags or drums and delivered to the site and warehouse prior to use, as shown in Figure 3.12. When needed, the virgin carbon is loaded into the adsorber charged via an adductor. Once the carbon becomes exhausted, it is slurried by gravity to a draining bin where the free water is removed and returned for treatment. The drained carbon is disposed of by landfill incineration. The advantage of this method is mainly in its simplicity. It is a sound system for low carbon exhaustion rate applications. However, the need to dispose of spent carbon is a definite drawback, especially when toxic hazardous materials are adsorbed on the carbon and it often becomes necessary to consider incineration of the spent carbon as the proper ultimate disposal.

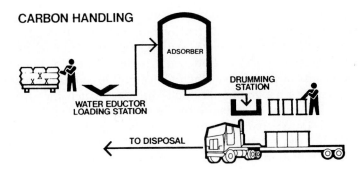

Figure 3.12 Throwaway carbon systems.

A second method shown in Figure 3-13 is custom off-site reactivation. This is somewhat similar to the throwaway approach from a carbon-handling standpoint, except that it assumes economies associated with carbon reuse. Spent carbon in this type of system is generally slurried by gravity from the adsorber to a draining bin where it is placed in drums and shipped to a firm providing reactivation on a custom basis. At the reactivation site, the carbon is unloaded and processed through a furnace. The dry-reactivated carbon is then placed back in drums and shipped back to the original plant. A bulk-reactivated carbon storage system is required because three separate carbon inventories are necessary: (1) spent carbon in the draining bin; (2) spent carbon in transit or at the reactivation site; and (3) reactivated carbon on site to be used when the carbon in the adsorber becomes exhausted.

The advantage of this approach is that it is generally more economical than throwaway carbon when the exhaustion rate falls in the range of 225–675 kg/day (500–1,500 lb/day). However, when compared to other reactivation alternatives, the number of handling steps and the high carbon attrition and losses are major disadvantages and the number of firms offering custom reactivation is limited.

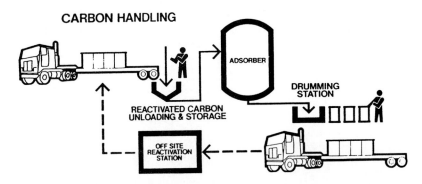

Figure 3.13 Off-site custom reactivation system.

Figure 3.14 is a typical granular activated carbon wastewater treatment system employing on-site reactivation. Wastewater is pumped at a controlled rate to the adsorbers where the organics are removed from the wastewater by adsorption. The treated water is discharged to the receiving stream or downstream processes. When the carbon becomes exhausted or the effluent reaches the maximum allowable discharge level, the spent carbon is transferred from the adsorber to the spent carbon storage tank. The spent carbon is then transferred to a furnace feed tank from which it is fed, at a constant rate, to a dewatering screw. This is an inclined screw conveyor, which drains off the transport water and provides a water seal for the top of the reactivation furnace or kiln.

The dewatered, and wet, carbon enters the furnace where the remaining moisture is evaporated (drying step). This is followed by destructive distillation of the adsorbed organics, which results in pyrolysis of a portion of the carbon from the organic materials (baking step). The carbon is then heated to the activation temperature where organic char is selectively destroyed, resulting in recovery of carbon activity (activation step). Drying occurs at a

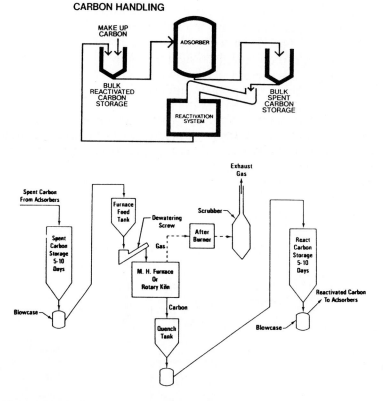

Figure 3.14   A. On-site reactivation system. B. Typical reactivation system flow diagram.

temperature of 100°C (212°F). The baking step occurs at about 650–760°C (1,200–1,400°F) and activation occurs at 870–1,000°C (1,600–1,800°F). Steam is added to the furnace or kiln, and the oxygen concentration is controlled to promote gasification of the fixed carbon, while minimizing the burning of the original granular activated carbon. Reactivated carbon exits the furnace and is cooled and wetted in the quench tank. This tank also serves as a bottom seal for the furnace. The reactivated carbon is then transferred to a reactivated carbon storage tank until needed.

One advantage of on-site reactivation is that by minimizing carbon handling, attrition, and spillage, carbon losses can be minimized. Also optimum conditions can be established in the furnace for maximizing the quality of reactivated carbon, while minimizing furnace oxidation losses. Drawbacks include the technological complexities of the reactivation system and the high capital cost. Furnace corrosion and maintenance problems are often severe, particularly in industrial wastewater applications. Air pollution problems are also always of concern.

A fourth spent carbon-handling alternative shown in Figure 3.15 is a complete adsorption service. Under this concept, the supplier designs, procures, and installs all required adsorption equipment. The service also includes replacement and reactivation of granular carbon as required to meet specific effluent objectives. System monitoring is included as part of the service. The supplier maintains the equipment and provides operators as an option.

Full service eliminates the need for any major capital investment by the user since the user is responsible only for site preparation, piping, and utilities to the battery limits of the system. Cost of the service may be accurately budgeted since it is determined in advance of system start-up. In this option, off-site control reactivation of spent carbon is employed. The user is not concerned with design or operation of reactivation facilities and their related problems. The service concept has application in the majority of municipal and industrial wastewater treatment cases where adsorption is the treatment process of choice. Carbon exhaustion rates ranging from as low as 45 kg/day

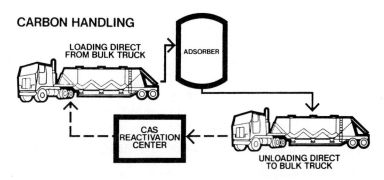

Figure 3.15 Full-service system.

(110 lb/day) to more than 14,000 kg/day (30,000 lb/day) can be handled economically using the full-service approach. The service circumvents concern
for inventories, carbon losses, labor, and design.

Contact Time with Carbon

In addition to the exhaustion rate, pilot testing should also establish the optimum superficial contact time to attain the desired level of organic removal.
In some cases, the contact time required for the minimum exhaustion rate
may be excessive, leading to high capital costs (Figure 3.16). Therefore, optimum contact time must be the balance of the capital and operating costs.
Contact times can range from as low as 7 minutes for taste and odor removal
to 20–50 minutes for sewage and industrial wastewater treatment applications. In some cases, contact times for removing complex organics from industrial wastes can be longer.

Type and Configuration of Adsorber

Adsorber configuration selected for a given application may depend on the
suspended solids in the influent wastewater, the shape of the adsorption
breakthrough curves, and the desired effluent quality. If the granular acti-

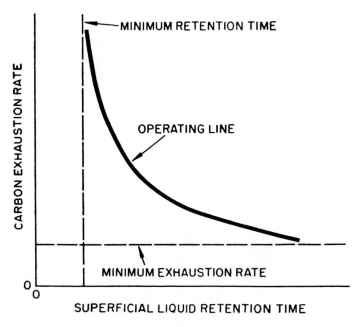

Figure 3.16 Relationship between exhaustion rate and contact time.

vated carbon is to be used as a filter to remove suspended solids as well as dissolved organics, then backwash capability must be built into the adsorbers. The need for suspended solids removal can be determined during the pilot testing program at the same time that contact time and breakthrough curve data are being collected. Most common adsorber designs include downflow-fixed beds, upflow-pulsed beds, and upflow-expanded beds. Adsorbers can be arranged in series or in parallel depending on the flow and treatment objectives. These configurations are illustrated in Figure 3.8.

Materials of Construction

Materials of construction for adsorbers, storage tanks, furnace internals, pumps, and slurry and process piping are important in the design and capital cost of a granular activated carbon adsorption system. Wet carbon in contact with carbon steel can result in rapid corrosive attack. Also, many wastewaters are corrosive in and of themselves. Proper selection of construction materials can reduce maintenance costs and help eliminate major outages.

Adsorbers and storage tanks are usually fabricated of lined carbon steel, stainless steel, FRP, and concrete. Lined steel is the most commonly used material of construction since the lining can be varied to suit the particular wastewater corrosion characteristics. Lined steel tanks are also generally less expensive than stainless steel. FRP tanks are competitively priced but are limited to atmospheric service. Common-wall concrete, deep-bed adsorbers have been used with limited success in several applications. Rubber, glass-reinforced polyester, vinyl ester, and modified epoxy linings have been used to line adsorbers. Corrosion resistance, abrasion resistance, shop versus field application, and cost should all be considered in selecting a lining material. Spray-on linings such as polyester and vinyl ester are less expensive than troweled-on modified epoxy linings and rubber linings; however, the latter two types of linings are generally more abrasion resistant.

Carbon slurry piping materials include carbon steel, stainless steel, and lined pipe, depending on the type of service. For noncorrosive fluids, Schedule 80 carbon steel is acceptable provided the lines are always flushed free of carbon at the end of a transfer. Schedule 40 stainless or lined steel pipe and valves are used with corrosive fluids such as those encountered in many industrial waste applications. Full-port stainless steel or FRP ball valves are recommended on all slurry lines to minimize carbon abrasion, head loss, and valve erosion.

Stainless steel, Alloy 20, ceramic-lined, and rubber-lined pumps are commonly used for pumping carbon slurries. Again, wastewater corrosivity, carbon abrasion, and pump erosion should be considered in selecting a pump. Stainless steel eductors and stainless steel or lined carbon steel blowcases are also used for transporting carbon slurries as shown in Figure 3-17.

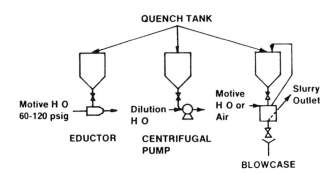

Figure 3.17 Carbon transfer equipment.

Stainless steel or Alloy 20 dewatering screws and quench tanks are generally used. Multihearth furnace arms and teeth are usually 25-12 stainless steel, although more exotic materials may be needed if the spent carbon is highly corrosive. The same is true for flighting in rotary kilns. Furnace brick is usually a standard high-alumina brick, although this can vary depending on the corrosivity of the spent carbon. Stack gas scrubbers are usually made of stainless steel, Alloy 20, or FRP.

Regeneration Furnaces

The reactivation furnace is further discussed in Chapter 7. On-site reactivation economically depends on many factors. A thorough analysis of each type of furnace, as well as of the reactivation characteristics of the carbon, is important in selecting a furnace. The exhaustion rate must be defined and the reactivation time, spent carbon quality and corrosivity, and reactivated carbon quality must be established. This is invariably done by a pilot testing program.

The residence time at reactivation temperature is a key factor in sizing a furnace. Lightly loaded carbons, such as those obtained from tertiary sewage treatment and water treatment applications, reactivate in less than 20 minutes. Carbon used to treat general industrial plant effluents and municipal secondary wastewater are typically reactivated in 20–40 minutes. Heavily loaded carbons from concentrated industrial waste streams may require as long as 60 minutes to reactivate. Therefore, the same exhaustion rate will not necessarily lead to the same size furnace for different applications.

Another factor affecting furnace sizing is the desired amount of excess capacity. If a plant expansion or process change is anticipated, additional capacity may be desired. In the case of municipal water and wastewater applications, the furnaces are usually designed with 50 percent excess capacity on an annual requirement basis to allow for intermittent or campaign-type operation. Industrial applications, on the other hand, are usually designed for continuous operation with built-in excess capacity. In this case, furnace turndown capability is important. Multihearth furnaces (Figure 3.18) and ro-

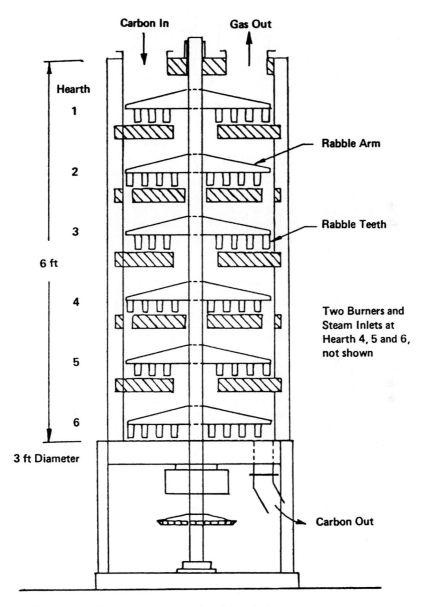

Figure 3.18 Cross-sectional view of multihearth furnace.

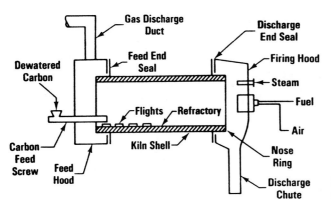

Figure 3.19 Cross-sectional view of a rotary kiln.

tary kilns (Figure 3.19) are both employed to reactivate granular activated carbons used in industrial and municipal wastewater treatment. Major factors influencing reactivation furnaces are:

- Capital cost, percent of multihearth cost.
- Reactivated carbon quality.
- Carbon losses—furnace only.
- Fuel consumption, percent of multihearth furnace.
- Corrosion and slagging resistance.
- Turndown potential.
- Space requirements.
- Degree of control.
- Maintenance.
- Effect of outages.
- Onstream factor.

Storage of Carbon

On-site carbon storage depends on the design philosophy of the wastewater treatment system. If continuous treatment is critical during maintenance and unscheduled outages, sufficient carbon inventory should be provided. Major maintenance shutdowns on a furnace generally require a minimum seven to ten days to complete. Sufficient spent carbon storage capacity and reactivated carbon inventory should be available. This inventory also provides flexibility in plant operation in that a relatively constant furnace feed rate can be maintained even if the exhaustion rate varies. Several types of bulk carbon storage systems are shown in Figure 3.20. In the case of throwaway carbon and off-site custom reactivation, facilities for spent carbon and reactivated carbon storage should also be provided. Factors affecting the amount of storage required include the exhaustion rate and the turnaround time to replace spent

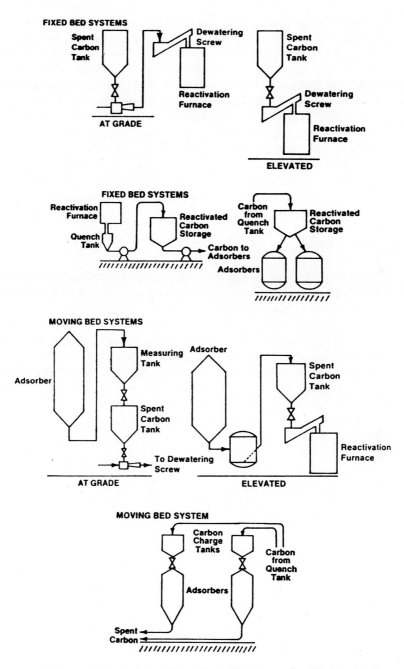

Figure 3.20 A. Spent carbon storage tank configurations—fixed-bed system; B. Reactivated carbon storage—fixed-bed systems; C. Spent carbon storage tank configurations—moving-bed systems; D. Reactivated carbon storage—moving-bed system.

carbon with reactivated or virgin carbon. In the case of off-site custom reactivation, sufficient reactivated carbon storage should be provided to allow continuous operation independent of the time required to drain, ship, reactivate, and return the reactivated carbon. Generally, virgin carbon is stored in drums or bags and spent carbon is stored in drums or bins. In the carbon-service approach, the supplier maintains an inventory of reactivated carbon at a central facility, thus minimizing the amount of on-site storage required. Also, the supplier operates his or her own shipping fleet to assure prompt delivery of the carbon as opposed to using common carriers.

The operating cost of granular activated carbon treatment systems is primarily a function of the carbon exhaustion rate. Once the exhaustion rate is known, the economics of various spent carbon-handling alternatives can be evaluated easily.

Carbon Losses

Makeup carbon costs represent the major operating expense for an on-site reactivation system. Makeup carbon can range from 5 percent to 10 percent of the daily exhaustion rate. Factors affecting carbon losses include type of carbon, carbon oxidation in the furnace, furnace outages, spillage, abrasion, and high gas flow rates in the furnace. A hard, dense carbon such as that made from bituminous coal is recommended for systems employing thermal reactivation to minimize abrasion and oxidation. A well-designed system will control slurry line velocities, backwash rates, gas flow rates, and overflow rates. Good plant housekeeping will also go a long way in minimizing losses. In general, a 7 percent makeup rate should be used for preliminary economic analysis.

4

Gas-Phase Adsorption and Air Pollution Control ═══

Adsorption for gas purification comes under the category of dynamic adsorption. Where a high separation efficiency is required, the adsorption would be stopped when the breakthrough point is reached. The relationship between adsorbate concentration in the gas stream and the solid may be determined experimentally and plotted in the form of isotherms. These are usually determined under static equilibrium conditions but dynamic adsorption conditions operating in gas purification bear little relationship to these results. Isotherms indicate the affinity of the adsorbent for the adsorbate but do not relate the contact time or the amount of adsorbent required to reduce the adsorbate from one concentration to another.

Factors which influence the service time of an adsorbent bed include the following:

- Grain size of the adsorbent.
- Depth of adsorbent bed.
- Gas velocity.
- Temperature of gas and adsorbent.
- Pressure of the gas stream.
- Concentration of the adsorbates.
- Concentration of other gas constituents which may be adsorbed at the same time.

- Moisture content of the gas and adsorbent.
- Concentration of substances which may polymerize or react with the adsorbent.
- Adsorptive capacity of the adsorbent for the adsorbate over the concentration range applicable over the filter or carbon bed.
- Efficiency of adsorbate removal required.

Complex factors enter into the process. For example, the carbon is vary rarely completely stripped of the adsorbed solvents during the regeneration process as this would require a very large quantity of steam to remove the last traces of solvent from the carbon (see Chapter 6). Also several solvents may be received simultaneously, each upsetting the equilibrium of the others. Such factors will tend to reduce the capacity of the carbon and lengthen the adsorption zone and, in order to maintain high recovery efficiency, deep-bed filters are generally used. There are no general relationships for calculating the capacity of an adsorber operating under dynamic conditions from a knowledge of the physical or chemical characteristics of the adsorbate or adsorbent.

In actual practice, the adsorptive capacity, height of bed, and stripping conditions for a given adsorption application are obtained experimentally in laboratory units when the design engineer does not have applicable experience available.

Dynamic adsorption tests should be conducted with gases related as closely as possible to the conditions under which the adsorption plant will operate. Whenever possible the tests should be on the gas stream to be treated containing components requiring removal.

Air pollution problems in which adsorption is considered a unit operation involve gaseous contaminants. The number of molecules present at the carbon surface is dependent on the number that reach the surface and on the residence time of these molecules on the carbon surface. If n molecules strike a unit area of a surface per unit time, and remain there for an average time, t, then σ number of molecules are present per unit area of surface:

$$\sigma = nt \tag{4-1}$$

Using cm^2 as unit surface and seconds as unit time, n is the number of molecules falling on 1 cm^2/sec. The number n thus denotes the number of molecules striking each cm^2 of the surface every second, and this number can be calculated using Maxwell's and the Boyle-Gay Lussac equations. The number n is directly related to the speed of the molecules within the system. It is important to realize that the velocity of the molecules is not dependent on the pressure of the gas, but the mean free path is inversely proportional to the pressure. Thus:

$$n = 3.52 \times 10^{22} \times \frac{p}{\sqrt{MT}} \tag{4-2}$$

where

p = pressure in Hg, mm
M = molecular weight
T = absolute temp., °K

From this equation at 20°C and 750 mm Hg pressure, the following values can be obtained:

$$H_2 : n = 11.0 \times 10^{23} \text{ molecules/cm}^2/\text{sec}$$

$$N_2 : n = 2.94 \times 10^{23} \text{ molecules/cm}^2/\text{sec}$$

$$O_2 : n = 2.75 \times 10^{23} \text{ molecules/cm}^2/\text{sec}$$

The molecule residence time t on the surface is difficult to determine as is the number n. Reflection experiments can indicate the residence time on a smooth surface because if a molecule is retained on the surface for any finite time, the angle of removal will be random.

Adsorption Forces

Forces causing adsorption are the same ones that cause cohesion in solids and liquids and are responsible for the deviation of real gases from the laws of ideal gases. Basic forces causing adsorption can be divided into two groups: intermolecular or van der Waals forces, and chemical forces, which generally involve electron transfer between the solid and the gas. Depending on which of these force types plays the major role in the adsorption process, we distinguish between physical adsorption, where van der Waals or molecular interaction forces are in prevalence, and chemisorption, where heteropolar or homopolar forces cause the surface interaction. In the process of adsorption when the individuality of the adsorbed molecule (adsorbate) and the surface (adsorbent) are preserved, we have physical adsorption. If, between the adsorbate and the adsorbent, any electron transfer or sharing occurs, or if the adsorbate breaks up into atoms or radicals bound separately, then we are presented with chemisorption.

While the theoretical difference between physical and chemical adsorption is clear in practice, the distinction is not as simple as it may seem. The following parameters can be used to evaluate an adsorbate-adsorbent system to establish the type of adsorption:

- Heat of physical adsorption is on the same order of magnitude as the heat of liquefaction, while the heat of chemisorption is of the corresponding chemical reaction. It has to be pointed out here that the heat of adsorption varies with surface coverage because of lateral interaction effects. Therefore, the heat of adsorption has to be compared on corresponding levels.
- Physical adsorption will occur under suitable temperature and pressure conditions in any gas-solid system, while chemisorption takes place only if the gas is capable of forming a chemical bond with the surface.

- A physically adsorbed molecule can be removed unchanged at a reduced pressure at the same temperature that the adsorption took place. The removal of the chemisorbed layer is far more difficult.
- Physical adsorption can involve the formation of multimolecular layers, while chemisorption is always completed by the formation of a monolayer. In some cases physical adsorption may take place on the top of a chemisorbed monolayer.
- Physical adsorption is instantaneous. It is the diffusion into porous adsorbents which is time consuming, while chemisorption may be instantaneous and generally requires activation energy.

Adsorption Rate

As discussed, the boundary layer is most important in the phase interaction. To achieve a high rate of adsorption, it is necessary to create the maximum obtainable surface area within the solid phase. High surface area is produced by creating a large number of microcapillaries in the solid. Commercial adsorbents such as activated carbon, silica gel, alumina, and so on, are prepared in this manner. While adsorption is nearly instantaneous, the passage of molecules through capillaries (pores) may involve some time. There have been observations that in rare cases it has taken several days to reach adsorption equilibrium. The movement of molecules into the pores is a diffusion process.

Equilibrium

Solid-gas phase interaction may occur under two conditions. One involves the random mixing of the phases; the other involves their direct relative motion. Thus, static adsorption occurs when the adsorption process takes place in relative rest, or random mechanical mixing of the phases of the solid-gas system takes place and ends in the establishment of an adsorption equilibrium among the interacting phases. Dynamic adsorption represents a sorption process accomplished under conditions of direct relative motion of one or both phases. In air pollution control most applications involve dynamic conditions. Where adsorption equilibrium is not reached, it is essential to survey the equilibrium conditions because their modified effect is of major importance in dynamic nonequilibrium systems.

Adsorption equilibrium is defined when the number of molecules arriving on the surface is equal to the number of molecules leaving the surface into the gas phase. The adsorbed molecules exchange energy with the structural atoms of the surface and, provided that the time of adsorption is long enough, they will be in a thermal equilibrium with the surface atoms. To leave the surface, the adsorbed molecule has to take up sufficient energy from the fluctuations of thermal energy at the surface so that the energy corresponding to the vertical component of its vibrations surpasses the holding limit.

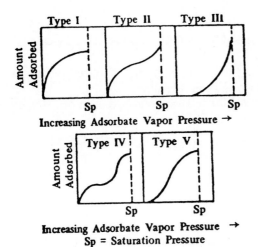

Figure 4.1 Types of adsorption isotherms as classified by Brunauer.

Isotherms, as measured under existing conditions, can yield qualitative information about the adsorption process and also indicate the fraction of the surface coverage, with certain assumptions to the surface area of the adsorbent.

In Figure 4.1, the five basic types of adsorption isotherms are presented as classified by Brunauer. The Type I isotherm represents systems in which adsorption does not proceed beyond the formation of a monomolecular layer. Such an isotherm is obtained when adsorbing oxygen on carbon black at $-183°C$. The Type II isotherm indicates an indefinite multilayer formation after the completion of the monolayer. As an example, the adsorption of water vapor on carbon black at 30°C results in such a curve. Type III isotherm is obtained when the amount of gas adsorbed increases without limit as its relative saturation approaches unity. The convex structure is caused by the heat of adsorption of the first layer becoming less than the heat of condensation due to molecular interaction in the monolayer. This type of isotherm is obtained when adsorbing bromine on silica gel at 20°C. The Type IV isotherm is a variation of Type II, but with a finite multilayer formation corresponding to complete filling of the capillaries. This type of isotherm is obtained by the adsorption of water vapor on active carbon at 30°C. The Type V isotherm is a similar variation of Type III obtained, for instance, when adsorbing water vapor on activated carbon at 100°C. Although a large number of equations have been developed to date based on theoretical considerations, none of them can be generalized to describe all systems.

Langmuir presented an ideal monolayer adsorption isotherm:

$$v = \frac{v_m b p}{1 + bp} \tag{4-3}$$

where

V = volume of gas (0°C, 760 mm Hg) adsorbed per unit mass of adsorbent

V_m = volume of gas (0°C, 760 mm Hg) adsorbed per unit of adsorbent with a layer one molecule thick

b = empirical constant in reciprocal pressure unit which has limited practical application

Brunauer expanded the Langmuir isotherm to include multilayer adsorption:

$$v = \frac{v_m C x}{(1 - x)[1 + (C - 1)x]} \tag{4-4}$$

where V_m and C are empirical constants and $x = P/p_S$. The constant C is derived from the heat of adsorption and V_m represents the volume of gas required to cover the surface with a monomolecular layer.

Although the BET equation has its limitations, such as the assumption that the heat of adsorption is constant over the entire surface coverage of the monolayer, and that the monolayer is completed before the formation of secondary layers with a heat of adsorption equaling that of the heat of liquefaction begins, it is very useful because it enables the numerical determination of surface area. Knowing the area occupied by a single molecule of adsorbent and the number of molecules needed to form a monolayer, it is possible to express the surface area of the adsorbent in cm²/g or m²/g.

Surface areas of commonly used adsorbents determined in this manner are:

Activated alumina	50–250 m²/g
Silica gel	200–600 m²/g
Molecular sieve	800–1,000 m²/g
Activated carbon	500–2,000 m²/g

In most gas-solid adsorption systems, the heat of adsorption is greater than the heat of evaporation or condensation of the same substance. This means that the entropy of the molecules when adsorbed on a particular surface will be greater than the entropy of the same molecules in their liquid or solid state. In studying gas-phase adsorption, as the van der Waal forces between different molecules are approximately the geometrical mean between the values for each of the two molecules, when combined with a molecule of its own kind, it is evident that the van der Waals forces of a gas molecule on the surface of a solid will be generally greater than the van der Waals forces holding it in liquid form. There are some exceptions to the fact that the heat of adsorption is higher than the heat of liquefaction. Such is the case, for instance, when water is adsorbed on activated carbon, the polar character of the water molecule causing only weak bonds. In this case, the heat of adsorption is indeed smaller than the heat of liquefaction. Adsorption never-theless takes place because the influence of the entropy difference is dominating. The fact that the entropy in the adsorbed state is higher than in a liquid state points to the fact that the adsorbed molecules have a greater degree of freedom than the molecules in the liquid state.

Dynamics of Adsorption

Adsorption applications in air pollution control generally involve the use of a dynamic system. The adsorbent is generally used in a fixed bed and con-taminated air is passed through the adsorbent bed. Depending on the con-centration and ergonomics, the contaminant is either recovered or discarded when the loading of the adsorbent requires regeneration. Although isotherms are indicative of the efficiency of an adsorbent for a particular adsorbate re-moval, they do not supply data to enable the calculation of contact time or the amount of adsorbent required to reduce the contaminant concentration below the required limits. Normal operation may be represented by Figure 4.2, which shows the building up of a saturated zone of adsorbers from the inlet end of the bed.

As more gas is passed through adsorption proceeds, the saturated zone moves forward until the breakthrough point is reached, at which time the exit concentration begins to rise rapidly above whatever limit has been fixed as the desirable maximum adsorbate level of the fluid. If the passage of the fluid is continued on still further, the exit concentration continues to rise until it becomes substantially the same as the inlet concentration. At this point, the bed is fully saturated. While the concentration when saturated is a func-tion of the material used and the temperature at which it is operated, the dynamic capacity is also dependent on the operating conditions, such as inlet concentration, fluid flow rate, and bed depth. The dependence of inlet con-

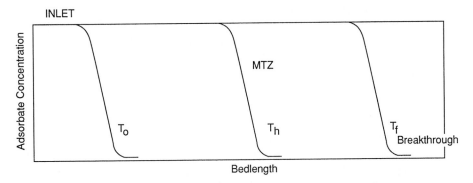

Figure 4.2 Formation and movement of the MTZ through an adsorbent bed: T_o = MTZ concentration gradient at the formation of the zone; T_h = MTZ con-centration gradient at half-life; and T_f = MTZ concentration gradient at break-through.

centration and fluid flow rate arise from heat effect and mass-transfer rates, but the dependence on bed depth, as can be seen from the preceding description, is dependent on the relative sizes of unsaturated and saturated zones. The zone of the bed where the concentration gradient is present is often called the mass-transfer zone (MTZ). Dynamic adsorption results are expressed in terms of the dynamic capacity, or breakthrough capacity at given inlet concentrations, temperatures, and flow-rate conditions of the bed, together with the bed dimensions. It is important that the adsorber bed should be at least as long as the transfer-zone length of the key component to be adsorbed. Therefore, it is necessary to know the depth of this mass-transfer zone.

Factors that play important roles in dynamic adsorption and the length and shape of the MTZ are:

- The type of adsorbent.
- The particle size of an adsorbent (may depend on maximum allowable pressure drop).
- The depth of the adsorbent bed and the gas velocity.
- The temperature of the gas stream and the adsorbent.
- The concentration of the contaminants to be removed.
- The concentration of the contaminants not to be removed, including moisture.
- The pressure of the system.
- The removal efficiency required.
- Possible decomposition or polymerization of contaminants on the adsorbent.

Adsorbent Selection

Most industrial adsorbents are capable of adsorbing both organic and inorganic gases. Preferential adsorption characteristics and other physical properties make each one more or less specific for a particular application. As an example, activated alumina, silica gel, and molecular sieves will adsorb water preferentially from a gas-phase mixture of water vapor and an organic contaminant. This is a drawback in the application of these adsorbents for organic contaminant removal. Activated carbon preferentially adsorbs nonpolar organic compounds. Silica gel and activated alumina are structurally weakened by contact with liquid droplets; therefore, direct steaming cannot be used for regeneration. (For steam regeneration see Chapter 6.)

In some cases, none of the adsorbents has sufficient retaining adsorption capacity for a particular contaminant. In such applications, a large surface area adsorbent is impregnated with inorganic or, in rare cases, with a high molecular weight organic compound, which can chemically react with the particular contaminant. Iodine-impregnated carbons have been used for removal of mercury vapor, bromine-impregnated carbons for ethylene or propylene removal. Action of these impregnants is either catalytic conversion or reaction to a nonobjectionable compound, or to a more easily adsorbed compound. It should be noted here that the general adsorption theory does not apply on the gross effects of the process. For example, the mercury removal by an iodine-impregnated carbon proceeds faster at a higher temperature, and a better overall efficiency can be obtained than at a low-temperature contact. An impregnated adsorbent is available for most compounds which, under particular conditions, are not easily adsorbed by nonimpregnated commercial adsorbents.

Adsorption takes place at the interphase boundary; therefore, the surface area of the adsorbent is an important factor in the adsorption process. Generally, the higher the surface area of the adsorbent, the higher is its adsorption capacity for all compounds. However, the surface area has to be available in a particular pore size within the adsorbent. At low partial pressure (concentration), the surface area in the smallest pores in which the adsorbate can enter is the most efficient. At higher pressures the larger pores are becoming more important, while at very high concentrations, capillary condensation will take place within the pores, and the total micropore volume is the limiting factor. Figure 4.3 shows the relationship between maximum effective pore size and concentration for the adsorption of benzene vapor at 20°C. The most valuable information concerning the adsorption capacity of a certain adsorbent is its surface area and pore-volume distribution curve in different diameter pores. Figure 4.4 shows the characteristic distribution

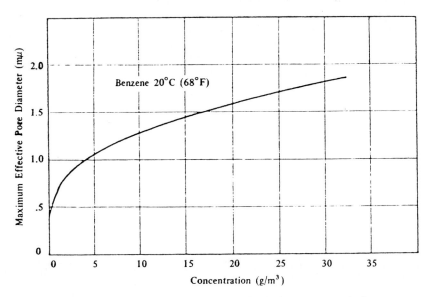

Figure 4.3 Relationship between pore size and vapor concentration.

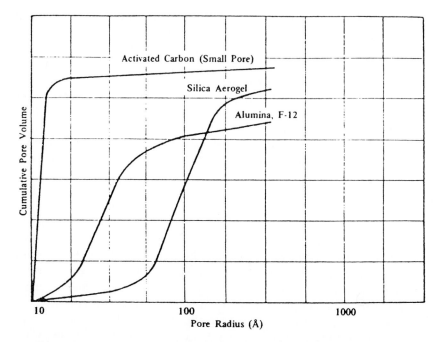

Figure 4.4 Cumulative pore volume versus pore size for different adsorbents.

curves for several different adsorbent types. As Figure 4.3 shows, the relationship between adsorption capacity and surface area in optimum pore sizes is concentration dependent and it is very important that any evaluation of adsorption capacity is performed under actual concentration conditions. In Figure 4.5 benzene adsorption isotherms are shown for several carbon types.

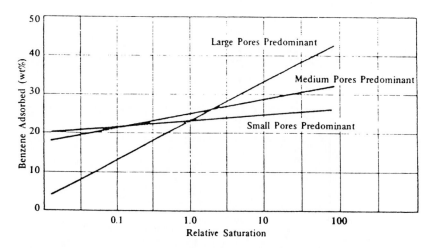

Figure 4.5 Benzene adsorption isotherms on various pore structure activated isotherms.

These lines cross at different concentrations, depending on the surface area distribution of the carbons.

The action of molecular sieves is slightly different from those of other adsorbents in that selectivity is determined more by the pore-size limitations of the molecular sieve. In selecting molecular sieves, it is important that the contaminant to be removed be smaller than the available pore size, while the carrier gas or the not-to-be-removed component is larger and thus is not absorbed. Because the optimum pore size varies with concentration, molecular sieves are limited in their use by the applicable concentration ranges. Figure 4.6 shows water adsorption isotherms for a molecular sieve, silica gel, and alumina.

Particle-Size Effects

The dimension and shape of particle size affect both the pressure drop through the adsorbent bed and the diffusion rate into the particles. The pressure drop is lowest when the adsorbent particles are spherical and uniform in size. The external mass transfer increases inversely with the $d^{3/2}$ and the internal adsorption rate inversely as d^2. The pressure drop will vary, with

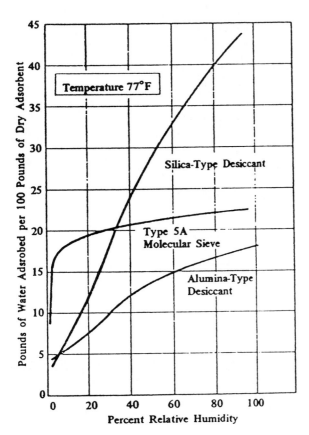

Figure 4.6 Water adsorption isotherms.

the Reynolds number being roughly proportional to velocity and inversely proportional to particle diameter. It is evident that everything else being equal, adsorbent beds consisting of smaller particles, although causing a higher pressure drop, will be more efficient. Therefore, sharper and smaller mass-transfer zones will be obtained.

Adsorbent Bed Depth

Bed depth on adsorption mass transfer has two effects. First, it is important that the bed be deeper than the length of the transfer zone which is unsaturated. The second is that any multiplication of the minimum bed depth gives more than a proportionally increased capacity. Generally, it is advantageous to size the adsorbent bed to the maximum length allowed by pressure-drop considerations. The determination of the depth of the MTZ or unsaturated depth may be determined experimentally:

$$\text{MTZ} = \frac{\text{Total bed depth}}{t_2/(t_2 - t_1) - x} \tag{4-5}$$

where

t_1 = time required to reach breakpoint
t_2 = time required to saturation
x = the degree of saturation in the MTZ

or,

$$\text{MTZ} = \frac{1}{1 - x} D_1 \left(1 - \frac{C_1}{C_s}\right) \tag{4-6}$$

where

D_1 = bed depth
C_1 = breakthrough capacity of bed D_1
C_s = saturation capacity
x = the degree of saturation in the MTZ

C_s of the preceding equation can be obtained by measuring the breakthrough capacities of two beds and using the following equation:

$$C_s = \frac{C_2 D_2 - C_1 D_1}{D_2 - D_1} \tag{4-7}$$

where

C_1 = breakthrough capacity for bed length of D_1
C_2 = breakthrough capacity for bed length of D_2

    Direct methods for the calculation of the MTZ are also possible using transfer units; however, particularly for multicomponent systems, the calculation becomes very complicated.

Gas Velocity

The velocity of the gas stream through adsorbent beds is limited by the adsorbent crushing velocity and varies with different types of adsorbents. The data on crushing velocities can be obtained from manufacturers of adsorbents. As an example, the crushing velocity for a 6 × 10 mesh nutshell carbon is:

$$V(MW)(P) < 50,000 \qquad (4\text{-}8)$$

where

$$V = \text{superficial velocity, ft/min}$$
$$MW = \text{molecular weight of gas}$$
$$P = \text{system pressure in atm}$$

Crushing velocity pressure is <50,000, determined experimentally. The length of the MTZ is directly proportional with velocity; at high velocities, the unsaturated zone is elongated.

Temperature Effects

As per basic adsorption theory, adsorption decreases with increasing temperature. Because the equilibrium capacity of adsorbent is lower at higher temperatures, the dynamic or break-through capacity will also be lower, and the MTZ is proportionally changed with temperature. In some cases, refrigerated systems are used to enhance or increase adsorption. The adsorption process is exothermic. As the adsorption front moves through the bed, a temperature front also proceeds in the same direction, and some of the heat is imparted to the gas stream. When the gas leaves the adsorption front, the heat exchange will reverse and the gas will impart heat to the bed. Increase in temperature during the adiabatic operation of the adsorber bed decreases the capacity of the adsorbent. The adiabatic temperature rise in an adsorber can be calculated by assuming that there is a thermal equilibrium between the gas and the bed, and that the temperature of the outlet gas stream is essentially the same as that of the bed. Increase in temperature during the adiabatic operation of the adsorber bed:

$$\Delta t = \frac{6.1}{(S_g/C) \times 10^5 + 0.51 \, (S_A/W)} \qquad (4\text{-}9)$$

where

$$t = \text{temperature rise, °F}$$
$$W = \text{saturation capacity of bed at } t + t, \text{ °F}$$
$$C = \text{inlet concentration, ppm}$$
$$S_g = \text{specific heat of gas, Btu/ft}^3/\text{°F}$$
$$S_A = \text{specific heat of adsorbent, Btu/lb/°F}$$

Values of S_A for common adsorbents are under ambient conditions:

Activated carbon	0.25
Alumina	0.21
Molecular sieve	0.25

Adsorbate Concentration

The adsorption capacity of adsorbents is directly proportional to the concentration of the adsorbate. The concentration of the adsorbate is inversely proportional to the length of the MTZ. Thus, all else being equal, a deeper bed will be required to remove a lower concentration contaminant with equal efficiency than to remove the same contaminant at higher concentrations. It is important that for combustible gases, the concentration entering the adsorbent be kept below the lower explosive limit. The concentration and value of the contaminants also determine if recovery of the adsorbate is justified.

Presence of Contaminants/Pressure

Some portion of all gases present will be adsorbed on the adsorbent surface. Because these gases compete for the available surface area and/or pore volume, their effect will be the lowering of the adsorption capacity for the particular adsorbate, which is to be removed. Under ambient conditions, very little (10–20 ml STP/g) air is adsorbed on commercial adsorbents; however, moisture or carbon dioxide has a more significant effect. Activated carbon is less sensitive to moisture than silica gel and alumina; at high gas-moisture content, its adsorption capacity can be considerably lower than adsorption from dry air stream. It is preferred to adsorb organic contaminants from the lowest relative humidity gas stream when using unimpregnated adsorbents. The reverse is true for most impregnated adsorbents, where the moisture enhances the reaction between the gaseous contaminants and the impregnating agent.

Adsorption capacity of an adsorbent increases with pressure, if the partial pressure of the contaminant increases. However, at high pressures (over 500 psig), a decrease in capacity will be observed due to retrograde condensation and a decrease in the fugacity of the more easily adsorbed compound and increased adsorption of the carrier gas.

At times it is sufficient to lower the adsorbate concentration only to a small extent, while in other cases, total removal is required. Deeper adsorbent beds are required to achieve a 99.9 percent single-pass removal which is better than a partial removal efficiency of 60 percent to 80 percent.

Some solvents or compounds may decompose, react, or polymerize when in contact with adsorbents. The decomposed product may be adsorbed at a lower capacity than the original substance or the decomposition product may have different corrosion or other properties. As an example, in an air stream, NO is converted to NO_2 when in contact with activated carbon. Polymerization on the adsorbent surface will significantly lower adsorption capacity and render it nonregenerable by conventional low-temperature methods such as steam.

An example is the adsorption of acetylene on activated carbon at higher temperatures. Decomposition may also take place in regenerative systems during direct steam stripping of the adsorbent bed. Factors influencing adsorption which are treated individually have a combined or interrelating effect on the adsorption system.

Dynamic adsorption in practice is a rather intricate process, influenced by a large number of complex factors. Some attempts have been made to develop a strictly theoretical formula for the design calculation of the adsorption system in practice adsorptive capacity, MTZ, and several other factors which should be experimentally determined in small-scale equipment.

Intermittent Operation

Often adsorbers are operated periodically or the concentration of the contaminant greatly varies depending on the periodic discharge of contaminants. The performance of the adsorption system is impaired under such conditions. This is caused by the variation of adsorbate concentrations with bed height. In Figure 4.7, an MTZ diagram of a system is shown where, under normal

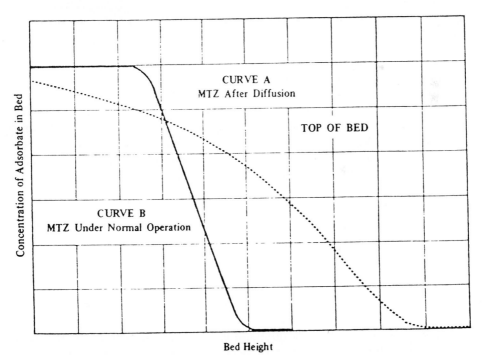

Figure 4.7 Effect of diffusion on MTZ.

operation, an MTZ curve (A) is obtained. The continued circulation of the carrier gas in the absence of contaminant causes the adsorbate to diffuse through the bed by the process of desorption into the carrier and readsorption until the low concentration causes elongation of the MTZ, represented by curve B (dashed line). Short periods of intermittent operation do not affect greatly the overall capacity of an adsorption system if the bed depth equals several MTZ lengths, but long periods of intermittent operation, particularly in an undersized system, will cause a serious capacity drop.

Regeneration

In regeneration of a system, the main factor—economics with in-place regeneration—is or is not preferred to the replacement of the entire adsorbent charge. It is also important to establish that the recovery of the contaminant is worthwhile, or if only the generation of the adsorbent is required. If recovery is the principal objective, the best design can be based on a prior experimental test to establish the ratio of the sorbent fluid to the recoverable adsorbent at the different working capacities of the adsorbent. A typical plant, for example, will have a steam consumption in the region of 1–4 lb of steam per lb of recovered solvent. Description of the contaminant can be achieved by several of the following different methods:

	Percentage of Charge Expelled
Heating at 100°C (212°F) for 20 min	15
Vacuum 50 mm Hg at 20°C (68°F) for 20 min	25
Gas circulation at 130°C (266°F) for 20 min	45
Direct steam at 100°C (212°F) for 20 min	98

Under most conditions, direct steam regeneration is the most efficient. The steam entering the adsorbent bed not only introduces heat, but adsorption and capillary condensation of the water will take place, which will supply additional heat and displacement for the desorption process. The following factors should be considered when designing the stripping process:

- Length of time required for the regeneration should be as short as possible. If continuous adsorption and recovery are required, multiple systems have to be installed.
- Short regeneration time requires a higher steaming rate, thus increasing the heat duty of the condenser system.
- Steaming direction should be in the opposite direction to the adsorption to prevent possible accumulation of polymerizable substances, and also to permit the shortest route for the desorbed contaminant.
- To enable a fast stripping and efficient heat transfer, it is necessary to sweep out the carrier gas from the adsorber and condenser systems as fast as possible.

- A larger fraction of the heat content of the steam is used up to heat the adsorber vessel and the adsorbent; thus, it is essential that the steam condenses quickly in the bed. The steam should contain only a slight superheat to allow condensation.
- It is advantageous to use a low-retentivity carbon to enable the adsorbate to be stripped out easily. When empirical data are not available, the following heat requirements have to be taken into consideration:
  - heat to the adsorbent and vessel
  - heat of adsorption and specific heat of adsorbate leaving the adsorbent
  - latent and specific heat of water vapor accompanying the adsorbate
  - heat in condensed, indirect steam
  - radiation and convection heat loss

Since the adsorbent bed must be heated in a relatively short time to reactivation temperature, it is necessary that the reactivation steam rate calculation is increased by some factor that will correct for the nonsteady-state heat transfer. During the steaming period, condensation and adsorption will take place in the adsorbent bed, increasing the moisture content of the adsorbent. A certain portion of the adsorbate will remain on the carbon. This fraction is generally referred to as *heel*. To achieve the minimum efficiency drop for the successive adsorbent cycles, the adsorbent bed should be dried and cooled before being returned to the adsorption cycle. The desired state of dryness will depend on the physical properties of the adsorbate and on the concentration of the adsorbate in the carrier stream. When using high-adsorbate concentrations, it may be desirable to leave some moisture in the adsorbent so that the heat of adsorption may be used in evaporating the moisture from the adsorbent, thus preventing any undue temperature rise of the adsorbent bed.

It is also necessary to establish the materials of construction on the basis that several compounds, especially chlorinated hydrocarbons, will undergo a partial decomposition during regeneration, forming hydrochloric acid.

Safety factors have to be considered in designing a regeneration system, assuring that the adsorber is not being used at temperatures higher than the self-ignition point of the contaminant. Carbon does not lower the ignition temperature of solvents and, as an example, solvent adsorbed on carbon ignites at the same temperature as the solvent vapor alone.

AIR POLLUTION CONTROL

Air pollution control is a broad and far-reaching subject that covers many complex unit operations, some consisting of filtration of fluid-solid particles, heat transfer and condensation, adsorption in a liquid state, and adsorption on a solid surface. When considering the problems of odor control and solvent recovery, which have become increasingly important in recent years due to the escalating costs, there are only two viable mechanisms: (1) oxidation by thermal or catalytic conversion, or (2) adsorption on one of several highly porous materials. The oxidation process is generally the last resort, used only where no other mechanism is possible, or where large quantities of waste heat may be recovered and/or used for processing purposes because of the very high energy requirements of heating the air stream as well as the small contaminant concentration in the gas stream. Adsorption has the dual advantage of generally having lower operating and capital costs as well as providing a means of solvent recovery in some processes.

ADSORPTION—THE PROCESS

Adsorption is a physical process that deals specifically with the concentration of dispersed materials in a continuous phase (carrier stream) on the surface of a highly porous material (Figure 4.8). From this figure, it can be seen that the rate of adsorption is influenced by diffusional resistances.

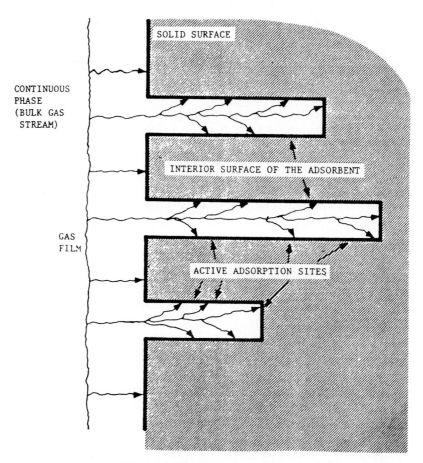

Figure 4.8 The adsorption process.

TABLE 4.1   HEATS OF PHYSICAL ADSORPTION AND CHEMISORPTION

System	Physical adsorption (cal/mole)	Chemisorption (cal/mole)	Heat of liquefaction
Nitrogen on Iron Catalyst	2,000–3,000	35,000	1,300
Oxygen on Charcoal	2,700 at 150 C	70,000 at room temperature, 200,000 at 400°C	
Hydrogen on Zinc Oxide	1,900 at 0°C	20,000 at 300–444°C	

Pollutants must first diffuse from the bulk gas stream across the gas film to the exterior surface of the adsorbent. Due to the highly porous nature of the adsorbent, the interior contains by far the majority of free surface area. For this reason, the molecule to be adsorbed must diffuse into the pore.

Surface Resistance

After diffusion into the pores, the molecules must then physically adhere to the surface, losing the bulk of their kinetic energy. The adhesion on the solid surface is due primarily to the imbalance within the structure of the adsorbent that creates a strong surface tension phenomenon. The molecules striking the surface lose their kinetic energy to the top layer of adsorbent molecules and thus satisfy an imbalance on the surface. Another consideration is that of van der Waals attraction in some cases.

Since adsorption is a physical process in which work is performed on the molecules, there must be associated with it some change in the thermodynamic state of the adsorbent during the adsorption process. The process is generally exothermic in nature, and the resultant evolution of heat, termed the *heat of adsorption*, is defined as the decrease in the heat content of the system. In general, the heat of adsorption is of the same magnitude as that of the heat of liquefaction. In some instances, the heat of adsorption will be much greater, and on the order of magnitude of the heat of reaction. This type of adsorption results from an activation of reactant molecules on the surface and is termed *activated adsorption of chemisorption*. Some typical heats of adsorption and chemisorption are shown in Table 4.1.

There have been many complex and highly theoretical approaches made to explain the phenomenon of adsorption. The problem is complex in nature because it is a function of not only physical parameters such as temperature and pressure, but also of concentration and molecular charge considerations. To bring the problem into a simple framework, the theories are presented on a broad base that will hold true only for specific materials and conditions. It can be stated, therefore, that the amount of a given gas adsorbed at equilibrium is a function of the final pressure and temperature only:

$$a = f(p, T) \tag{4-10}$$

The usual means of presenting adsorption data as has been discussed is by use of the adsorption isotherm, which is a plot of the amount adsorbed versus the pressure (or concentration if a gas) at constant temperature. Not all isotherms have the same shape. The theories of adsorption are complex, with many empirically determined constants. For this reason, pilot data should always be obtained on the specific pollutant adsorbent combination prior to full-scale engineering design.

SYSTEMS

Adsorption systems used for odor control or solvent recovery are of two types—regenerative and nonregenerative. Some of the requirements for systems design are:

- Long enough duration of contact (detention time) between air stream and sorbent bed for adequate sorption efficiency.
- Sufficient sorption capacity to provide the desired service life.
- Small enough resistance to air flow to allow adequate operation of air-moving devices being used.
- Uniformity of distribution of air flow over the sorbent bed to ensure full utilization of the sorbent.
- Adequate pretreatment of the air to remove nonadsorbable particles which would impair the action of the sorbent bed.
- Provision for renewing the sorbent after it has reached saturation.

To establish the required contact time in the adsorber, it is necessary to obtain a plot of vapor concentration versus depth of bed (Figure 4.9). This plot should be obtained experimentally or from a reliable sorbent manufacturer who has had experience with the particular system. As can readily be seen by the shape of the vapor concentration curve, this is not a linear function and should not be assumed to be. The entire curve shifts to the right as a function of time (or amount of vapor passing through the unit). When the exit concentration is equal to the threshold concentration ($C_e = C_t$), this is defined as the *breakthrough point*. The depth of the bed from the point where the vapor concentration equals the inlet concentration ($C_v = C_t$) to the breakthrough point is termed the *minimum-transfer zone* and is the absolute minimum depth of bed permissible. The overall depth of bed is normally dependent on the sorptive capacity of the adsorbent, which can be obtained from the adsorption isotherm of the system.

The air resistance, or pressure drop, across the system is a function of the sorbent particle size, the required bed depth, and the superficial linear velocity through the bed. The range of air flows through commercial adsorption units is 25 ft/min for thin-bed units to 80 ft/min in thick-bed units.

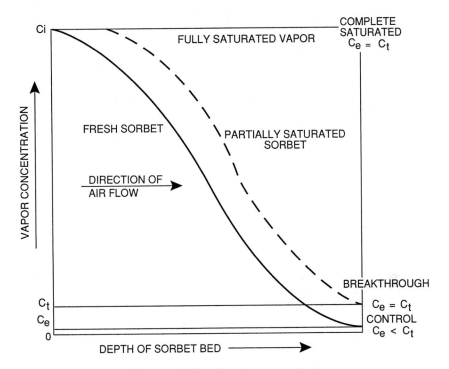

Figure 4.9 The adsorption wave front.

A typical plot of pressure drop versus superficial linear velocity is shown in Figure 4.10.

In many air pollution control applications, the problem may not simply be one of odor control or solvent recovery. If there are particulate materials in the air stream, the adsorbent bed may become fouled or plugged, thereby increasing the system pressure drop and possibly decreasing the adsorption efficiency of the bed. In such cases, a precleaner is required. Normally dry collection techniques such as filtration, cyclone separation, or use of a dry electrostatic precipitator is preferred. The use of wet collection methods is generally undesirable due to the increase in relative humidity of the carrier stream. In cases in which a scrubber must be used, a chiller and reheater can be used to lower the relative humidity of the influent to the adsorption system. Small quantities of moisture actually enhance the adsorption process, as the heat of adsorption is carried with the moisture; however, relative humidities in excess of 50 percent tend to lessen the effectiveness of the bed.

Precleaning would be required where fine metallic particulate is part of the effluent to be handled. The metallic particulate may impregnate the surface of the carbon or other adsorbent and act as a catalyst to oxidize some organic materials. The danger here is twofold; the heat of adsorption may raise the temperature of the bed to an unsafe level and solvent recovery would not yield sufficiently pure products.

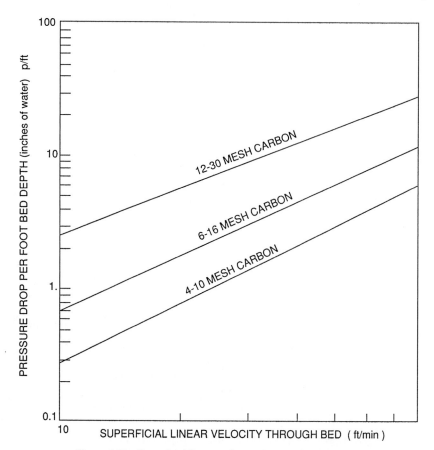

Figure 4.10 Superficial linear velocity through bed (ft/min).

Under normal operation, the adsorption unit will continue in operation until just before the breakthrough point is reached. Generally, there is no instrumentation installed on the unit, and the time of breakthrough is determined by experience. In some cases, where valuable materials are being recovered or where the vapor is toxic, hydrocarbon or specific compound monitors may be used so that there is no loss of material. When the unit has reached breakthrough, the carbon must be either replaced or regenerated. Regeneration may be performed in situ or returned to the manufacturer. In situ regeneration is most commonly used in large, thick-bed units by means of steam injection to the bed in a counterflow direction to that of the air stream. The amount of steam required depends on the molecular weight of the sorbed material and is specified as pounds of steam per pound of solvent. Complete desorption of material is not economically or thermodynamically possible using steam regeneration. The higher the steam temperature, the greater degree of desorption that will be attained. After repeated regenerations, the breakthrough time may begin to show some sign of reducing, possibly because of material polymerizing on the adsorbent. The bed should be removed and returned to the manufacturer for thermal regeneration and reactivation, or credit may be given on new adsorbent.

In solvent recovery systems, the steam from the regeneration is condensed and separated by use of flash separators, settling tanks, or distillation units. In cases in which multiple solvents are used, fractional distillation may be required.

For the overall adsorption system to operate continuously, provision must be made for a second unit to be used during the regeneration of the initial bed. The total system will consist of two or more adsorption units that can be run either in parallel or in series to facilitate regeneration and cooling during the operation of one unit. A typical thick-bed regenerative system with solvent recovery is shown in Figure 4.11.

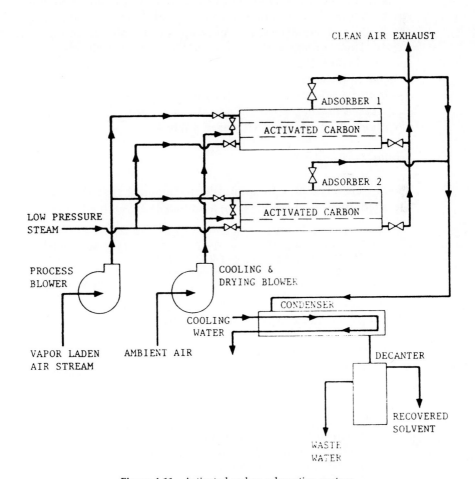

Figure 4.11 Activated carbon adsorption system.

Activated Carbon

The use of carbon as an adsorbent dates back to 1773 when Scheele described experiments on gases exposed to carbon. Various other experiments were carried out using carbon to create high vacuums. One of the earlier uses of carbon, in the form of wood charcoal, was for respirators and gas masks. It was found that charcoal made from different types of wood exhibited marked differences in the adsorptive capacities.

The process of activation is a slow dehydration and carbonization that is usually effected by heating the raw material in the absence of air. Experimental work has shown that activation of carbon can only occur on chars that have been made at temperatures below 600–700°C. The activation process selectively enlarges the pores of the carbon to provide high adsorptive capacity. The surface of activated carbon presents a largely homogeneous distribution of electrical charge. For this reason, activated carbon does not show any preferential adsorption of polar molecules such as water, but rather will desorb polar materials in favor of nonpolar or materials of higher molecular weight. Thus, a nonpolar solvent would be preferentially adsorbed in a water-solvent mixture. Because there are many types and grades of activated carbon available commercially, it has become necessary to establish specifications to obtain the proper carbon. The activity and retentivity of the carbon are generally based on their ability to adsorb a standard CCl_4 solvent. A typical set of carbon specifications is shown in Table 4.2.

Zeolites (Molecular Sieves)

Zeolites are crystalline aluminosilicate minerals first discovered by Baron Cronstedt in 1756. A zeolite is an aluminosilicate with a framework structure

TABLE 4.2 TYPICAL SPECIFICATIONS FOR ACTIVATED CARBON USED FOR AIR PURIFICATION

Property	Specification
Activity for CCl_4 [a]	At least 50%
Retentivity for CCl_4 [b]	At least 30%
Apparent Density	At least 0.4 g/ml
Hardness (ball abrasion)[c]	At least 80%
Mesh Distribution	6–14 range (Tyler sieve series)

[a] Maximum saturation of carbon at 20°C and 760 torr in an air stream equilibrated with CCl_4 at 0°C.

[b] Maximum weight of adsorbed CCl_4 retained by carbon on exposure to pure air at 20°C and 760 torr.[4]

[c] Percent of 6–8 mesh carbon which remains on a 14-mesh screen after shaking with 30 steel balls of 0.25–0.27-in. diameter per 50 g carbon, for 30 minutes in a vibrating or tapping machine.

endorsing cavities occupied by large ions and water molecules, both of which have considerable freedom of movement, permitting ion exchange and reversible dehydration. Activation of zeolites is a dehydration process accomplished by the application of heat in a high vacuum. Some zeolite crystals show behavior opposite to that of activated carbon in that they selectively adsorb water in the presence of nonpolar solvents. Zeolites can be made to have specific pore sizes that will increase their selective nature due to the size and orientation of the molecules to be adsorbed. Molecules above a specific size could not enter the pores and therefore would not be adsorbed.

ENGINEERING DESIGN

To effect the good engineering design of an activated carbon adsorption system, it is first necessary to obtain the following data:

- The actual cubic feet per minute (ACFM) of air to be processed by the adsorber.
- The temperature of gas stream.
- The material(s) to be absorbed.
- The concentration of the material to be adsorbed.
- The odor threshold of the material to be adsorbed (Table 4.3).

TABLE 4.3   ODOR THRESHOLD CONCENTRATIONS

Substance	ppm
Carbon Tetrachloride	71.8
Ammonia	53.0
Phosgene	5.6
Chlorine	3.5
Acrolein	1.8
Amyl Acetate	1.0
Pyridine	0.23
Hydrogen Sulfide	0.18
Oil of Wintergreen	0.066
Crotonaldehyde	0.062
Benzyl Sulfide	0.006
Diphenyl Ether	0.0012
Isoamyl Mercaptan	0.00043
Ethyl Mercaptan	0.00026
Vanillin	0.000079
Butyric Acid	0.000065
Artificial Musk	0.0000034

- The presence of other pollutants in the air stream.
- Is solvent recovery required or justified?

After this information has been obtained, the cyclic time of the system must be determined. This is primarily an economic consideration, and should be reviewed and reevaluated after the initial sizing of the system. Should the initial capital cost of the adsorber be too high, the cyclic time may be reduced to enable use of a smaller system. In general, the larger the system, the greater the overall efficiency, and the less energy that will be spent per pound of material adsorbed. The normal starting point would be to choose a half working shift cyclic time so that the unit changeover could be made during a working break.

The weight of adsorbent required is then determined using the following equation:

$$W = \frac{t\, e\, Q_r M\, C_v}{6.43(10)^6\, S} \tag{4-11}$$

where

$\quad t =$ duration of adsorbent service before saturation (hr)
$\quad e =$ sorption efficiency (fractional)
$\quad Q_r =$ air flow rate through the sorbent bed (ACM)
$\quad M =$ average molecular weight of the sorbed vapor
$\quad C_v =$ entering vapor concentration (ppm by volume)
$\quad S =$ proportionate saturation of sorbent (fractional)

Refer to Table 4.4 for typical maximum values (retentivities).

The sorption efficiency e is a variable determined by the characteristics of the particular system, including concentration and temperature. For the purposes of engineering design calculations, it is normally assumed to be 1.0. The design engineer must also control the inlet temperature to be less than 100°F at the inlet to the unit.

The next step is to calculate the volume of carbon required based on the bulk density of the carbon (D_c):

$$V_{\text{bed}} = \frac{W}{D_c} \tag{4-12}$$

An equation for the overall pressure drop of the system is determined as follows:

$$\text{Area of bed} \quad\quad A_b = \frac{Q_I}{V_S} \tag{4-13}$$

TABLE 4.4 RETENTIVITY OF VAPORS BY ACTIVATED CARBON (percent retained in a dry air stream at 20°C, 760 m by weight)

Substance	Retentivity (%)	Remarks
Acetaldehyde	7	Reagent
Acetic Acid	30	Reagent, sour vinegar
Acetone	15	Solvent
Acetylene	2	Welding and cutting
Acryaldehyde	15	Acrolein, burning fats
Acrylic Acid	20	
Ammonia	Negligible	
Amyl Acetate	34	Lacquer solvent
Amyl Alcohol	35	Fuel oil
Benzene	24	Benzol, paint solvent, and remover
Body odors	High	
Bromine	40 (dry)	
Butane	8	Heating gas
Butyl Acetate	28	Lacquer solvent
Butyl Alcohol	30	Solvent
Butyl Chloride	25	Solvent
Butyl Ether	20	Solvent
Butylene	8	
Butyne	8	
Butyraldehyde	21	Present in internal combustion exhaust, i.e., diesel
Butyric Acid	35	Sweat, body odor
Camphor	20	
Caprylic Acid	35	Animal odor
Carbon Disulfide	15	
Carbon Tetrachloride	45	Solvent, cleaning fluid
Chlorine	15 (dry)	
Chloroform	40	Solvent, anesthetic
Cooking Odors	High	
Cresol	30	Wood preservative
Crotonaldehyde	30	Solvent, tear gas
Decane	25	Ingredient of kerosene
Diethyl Ketone	30	Solvent
Essential Oils	High	
Ethyl Acetate	19	Lacquer solvent
Ethyl Alcohol	21	Grain alcohol
Ethyl Chloride	12	Refrigerant, anesthetic
Ethyl Ether	15	Medical ether, reagent
Ethyl Mercaptan	23	Garlic, onion, sewer
Ethylene	3	More retentivity by reaction

(continued)

TABLE 4.4　RETENTIVITY OF VAPORS BY ACTIVATED CARBON (percent retained in a dry air stream at 20°C, 760 m by weight) (*continued*)

Substance	Retentivity (%)	Remarks
Eucalyptole	20	
Food (raw) odors	High	
Formaldehyde	Negligible	Disinfectant, plastic ingredient
Formic Acid	7	Reagent
Heptane	23	Ingredient of gasoline
Hexane	16	Ingredient of gasoline
Hydrogen Bromide	12	
Hydrogen Chloride	12	
Hydrogen Fluoride	10	
Hydrogen Iodine	15	
Hydrogen Sulfide	3	Oxidizes to increase retentivity considerably
Indole	25	In excreta
Iodine	40	
Iodoform	30	Antiseptic
Isopropyl Acetate	23	Lacquer solvent
Isopropyl Alcohol	26	Solvent
Isopropyl Chloride	20	
Isopropyl Ether	18	Solvent
Menthol	20	
Methyl Acetate	16	Solvent
Methyl Alcohol	10	Wood alcohol
Methyl Chloride	5	Refrigerant
Methyl Ether	10	
Methyl Ethyl Ketone	25	Solvent
Methyl Isobutyl Ketone	30	Solvent
Methyl Mercaptan	20	
Methylene Chloride	25	
Naphthalene	30	Reagent, moth balls
Nicotine	25	Tobacco
Nitric Acid	20	
Nitro Benzene	20	Oil of bitter almonds Oil of mirbane
Nitrogen Dioxide	10	Hydrolyzes to increase retentivity
Nonane	25	Ingredient of kerosene
Octane	25	Ingredient of gasoline
Ozone	Decomposes to oxygen	Generated by electrical discharge
Packing-House Odors	Good	
Palmitic Acid	35	Palm oil
Pentane	12	Light naphtha
Pentylent	12	

TABLE 4.4 RETENTIVITY OF VAPORS BY ACTIVATED CARBON (percent retained in a dry air stream at 20°C, 760 m by weight) *(continued)*

Substance	Retentivity (%)	Remarks
Phenol	30	Carbolic acid, plastic ingredient
Propane	5	Heating gas
Propionic Acid	30	
Propylene	5	Coal gas
Propyl Mercaptan	25	
Propyne	5	
Putrascine	25	Decaying flesh
Pyridine	25	Burning tobacco
Sewer odors	High	
Skatole	25	In excreta
Sulfur Dioxide (dry)	10	Oxidizes to sulfur trioxide, common in city atmospheres
Sulfur Trioxide	15	Hydrolyzes to sulfuric acid
Sulfuric Acid	30	
Toilet Odors	High	
Toluene	29	Manufacture of TNT
Turpentine	32	Solvent
Valeric Acid	35	Sweat, body odor, cheese
Water	None	
Xylene	34	Solvent

where V_s is the superficial linear velocity through the bed.

$$\text{Height of bed } H_p = \frac{V_{bed}}{A_b} \tag{4-14}$$

$$\text{Total pressure drop } \Delta P_T = H_b \, (\Delta p/ft) \tag{4-15}$$

where p/ft is obtained from Figure 4.10.
Substituting Equations 4-13, 4-14, and 4-15 and rearranging yields:

$$\Delta P_T = \frac{V_{bed}\, V_s}{Q_r} \, (\Delta p/ft) \tag{4-16}$$

The final system pressure drop is then determined by an economic balance between the size tank required for a specific velocity and the power requirements for the pressure drop. Once the pressure drop has been determined, the height and area of the bed are calculated using Equations 4-5 and 4-6, and a cylindrical tank or pressure vessel of the appropriate size is selected (Figure 4.11). If the system is to be used for solvent recovery, the

TABLE 4.5 PROPERTIES OF GASES AND VAPORS AND THEIR RETENTION BY ACTIVATED CARBON

Substance	Formula	Molecular weight	Boiling point 760 mm C	Critical temperature °C	Approximate retentivity in % at 20°C 760 mm
Propionaldehyde	C_2H_5CHO	58.08	48.8		14
Butyraldehyde	C_3H_7CHO	72.10	75.7		21
Valericaldehyde	C_4H_9CHO	86.13	103.4		28
Acrylaldehyde	C_2H_3CHO	56.06	52.5		15
Crotonaldehyde	C_3H_5CHO	70.09	104.0		30
Formic Acid	$H.COOH$	46.03	100.7		7
Acetic Acid	CH_3COOH	60.05	118.1	321.6	40
Propionic Acid	C_2H_5COOH	74.08	141.1	339.5	40
Butyric Acid	C_3H_7COOH	88.10	163.5	335.0	40
Valeric Acid	C_4H_9COOH	102.13	187.0	379.0	40
Acrylic Acid	C_2H_2COOH	76.06	141.9		
Caprylic Acid	$C_7H_{15}COOH$	144.21	237.5		25
Pamitic Acid	$C_{15}H_{31}COOH$	256.42	339.0		25
Lactic Acid	$CH_3CHOH.COOH$	90.08	122.0		15
Methyl Acetate	$CH_3COO.CH_3$	74.08	57.1	233.7	16
Ethyl Acetate	$CH_3COO.C_2H_5$	88.10	77.15	250.1	19
Propyl Acetate	$CH_3COO.C_3H_7$	102.13	101.6	276.2	23
Butyl Acetate	$CH_3COO.C_4H_9$	116.16	126.5	288.0	28
Amyl Acetate	$CH_3COO.C_5H_{11}$	130.18	148.0	326.0	34
Acetone	$CH_3CO.CH_3$	58.08	56.5	235.0	15
Methyl Ethyl Ketone	$CH_3CO.C_2H_5$	72.10	79.6		25
Diethyl Ketone	$C_2H_5CO.C_2H_5$	86.13	102.7		30
Dipropyl Ketone	$C_3H_7CO.C_3H_7$	114.18	144.0		35
Methyl Ether	$(CH_3)_2O$	46.07	−23.6	10	
Ethyl Ether	$(C_2H_5)_2O$	74.12	34.6	192.8	15
Propyl Ether	$(C_3H_7)_2O$	102.17	91.0		18
Butyl Ether	$(C_4H_9)_2O$	130.23	142.0		20
Amyl Ether	$(C_5H_{11})_2O$	158.28	190.0		20
Methyl Acrylate	$C_3H_3O_2CH_3$	86.09	80.5		High
Ethyl Acrylate	$C_3H_3O_2C_2H_5$	100.11	99.8		High
Methyl Mercaptan	CH_3SH	48.10	7.6	196.8	20
Ethyl Mercaptan	C_2H_5SH	63.13	34.7	225.5	23
Propyl Mercaptan	C_3H_7SH	76.15	68.0		25
Ecalyptol	$C_{10}H_{18}O$	154.25	176.0		20
Camphor	$C_{10}H_{16}$	152.23	204.0		20
All essential oils					High
Methyl Chloride	CH_3Cl	50.49	−24.22	143.1	15

TABLE 4.5   PROPERTIES OF GASES AND VAPORS AND THEIR RETENTION
BY ACTIVATED CARBON (*continued*)

Substance	Formula	Molecular weight	Boiling point 760 mm C	Critical temperature °C	Approximate retentivity in % at 20°C 760 mm
Ethyl Chloride	C_2H_5Cl	64.52	12.2	187.2	20
Propyl Chloride	C_3H_7Cl	78.54	47.2	230.0	25
Butyl Chloride	C_4H_9Cl	92.57	78.0		30
Methylene Chloride	CH_2Cl_2	84.94	40.1		30
Chloroform	$CH.Cl_3$	119.39	61.26	263.0	40
Carbon Tetrachloride	CCl_4	153.84	76.0	283.1	45
Iodoform	CH_{13}	393.78	—		30
Phosgene	$COCl_2$	98.92	8.3	183.0	20
Pyridine	C_5H_5N	79.10	115.3	344.0	32
Indole	C_8H_7N	117.14	254.0		25
Skatole	C_9H_9N	131.17	266.2		25
Nicotine	$C_{10}H_{14}N$	162.23	247.3		25+
Nitrobenzene	$C_6H_5NO_2$	123.11	210.9		20
Urea	$CO(NO_2)$	60.06	Decomposes		15
Uric Acid	$C_5H_4N_4O_3$	168.11	Decomposes		20
Putrescine	$(CH_2)_4(NH_2)_2$	88.15	158.0		25
Packing House Odors		Nitrogen compounds			
Cooking Odors		Nitrogen and Sulphur compounds			
Sewer Odors		Nitrogen and Sulphur compounds			
Chlorine	C_{12}	70.91	−33.7	144.0	15
Bromine	Br_2	159.83	58.78	302.0	40
	I_2	253.84	183.0	553.0	40+
Hydrogen Fluoride (Hydrofluoric Acid)	HF	20.01	19.4	4	
Hydrogen Chloride (Hydrochloric Acid)	HCl	36.47	−83.7	51.4	5
Hydrogen Bromide	HBr	80.92	−67.0	90.0	8
Hydrogen Iodide	HI	127.93	−35.38	151.0	10
Nitrogen Dioxide	NO_2	46.01	21.3	158.0	5
(Nitrogen Tetraoxide)	(N_2O_4)	(92.02)			
Nitric Acid	HNO_3	63.02	5		
Sulfur Dioxide	SO_2	64.06	−10.0	157.2	10 dry
Sulfur Trioxide	SO_3	80.06	44.8	218.3	15 dry
Sulfuric Acid	H_2SO_4	98.08	330.0	20	
Hydrogen Sulphite	H_2S	34.08	−61.8	100.4	3 dry
Water	H_2O	18.02	100.0	374.0	None

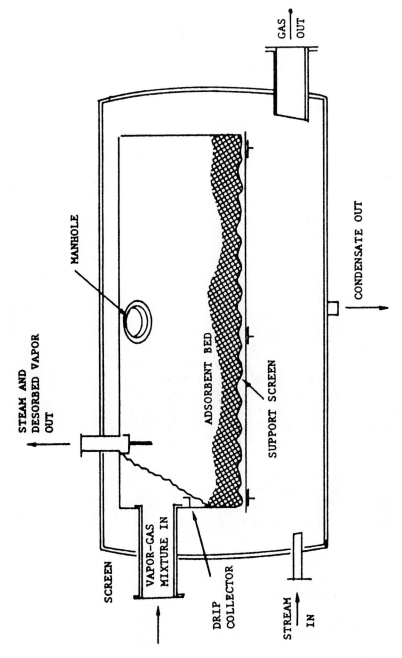

Figure 4.12 Typical activated carbon-bed adsorber.

STEAM AND DESORBED VAPOR OUT

MANHOLE

SCREEN

VAPOR–GAS MIXTURE IN

DRIP COLLECTOR

STREAM IN

ADSORBENT BED

SUPPORT SCREEN

CONDENSATE OUT

GAS OUT

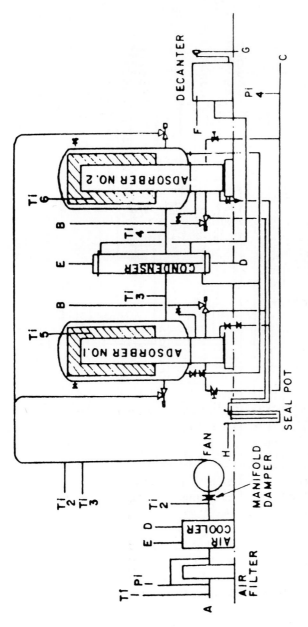

A – SOLVENT LADEN AIR INLET
B – STRIPPED AIR OUTLET
C – STEAM INLET
D – COOLING WATER SUPPLY

E – COOLING WATER RETURN
F – RECOVERED SOLVENT OUTLET
G – EFFLUENT TO DRAIN
H – CONDENSATE TO DRAIN

Figure 4.13 Preliminary flow sheet for solvent recovery plant.

111

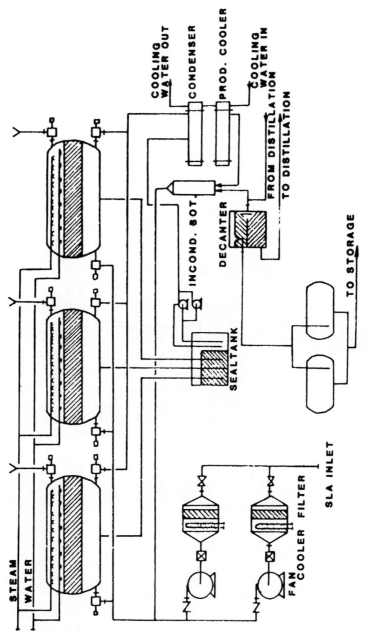

Figure 4.14 Adsorption system.

TABLE 4.6 TYPICAL APPLICATIONS AND EFFICIENCY OF CARBON ADSORPTION SYSTEMS

Compound	Control efficiency (%)	Comments
Acetone/phenol	92	Overall hydrocarbon removal efficiency
	83.4	Efficiency calculated from design data
	99	Efficiency including condenser
Dimethyl	80	VOC removal efficiency
terephrhalate	97	p-Xylene removal efficiency
Maleic anhydride	85	System control efficiency
Methylene chloride	>90	Reported efficiency for controlling emission from pharmaceutical manufacturing
Perchloroethylene	96, 99	Perchloroethylene control efficiency
	96, 97	Test data from dry-cleaning industry

amount of solvent retained in the bed must be calculated to determine how much steam to use for stripping, where:

$$\text{lb of solvent} = S\,W$$

Solvent recovery systems would also necessitate the specification of condenser duties, distillation tower sizes, holding tanks, piping, and valves.

In summary, engineering design of an adsorption system should be based on pilot data for the particular system. Information can usually be obtained directly from the adsorbent manufacturer. The overall size of the unit is determined primarily by economic considerations, balancing the operating costs against the capital costs. Adsorption, as can readily be seen, is not an exact science, but rather an art that draws on the experience of the design engineer. Figures 4.12, 4.13, and 4.14 are schematic diagrams of air pollution/solvent recovery systems.

SOLVENT RECOVERY

Volatile solvents vaporized during a manufacturing process may be recovered and used again. From the mixture of air and vapor, which is generally the form in which the solvent must be sought, the latter may be condensed to a liquid and trapped by the application of cold and moderate pressure; the vapor-laden air may be passed through a liquid adsorbent such as water; or finally, the mixture may be passed through a sufficiently thick bed of a solid adsorbent such as activated carbon and later driven off by steam. There are certain conditions which each of these processes meets better than the other two. Condensation by cooling may be properly selected when the concentration of the vapor is very high; adsorption in oil in a long series of plate towers has been the general practice for separating natural gasoline from hydrocarbon gases; in the general chemical and allied industries, adsorption on activated carbon is favored.

An installation for continuous operation consists of a blower, two adsorbers, a condenser, a decanting vessel to separate solvents which are immiscible with water, and more or less elaborate rectification or distillation equipment for solvents which are miscible with water. The adsorbers are usually built of steel, and may be lagged or left unlagged; the horizontal type is shown in Figure 4.15. The vapor-laden air is fed by the blower into one adsorber which contains a bed of 6- to 8-mesh activated carbon granules 12 to 30 inches thick. The air velocity through the bed is 40 to 90 feet per minute. The carbon particles retain the vapor; only the denuded air reaches the exit, and then the exhaust line. The adsorption is allowed to continue until the carbon is saturated, when the vapor-laden air is diverted to the second adsorber, while the first adsorber receives low-pressure steam fed in below the carbon bed. The vapor is reformed and carried out by the steam. The two are condensed and if the solvent is not miscible with water, it may be decanted continuously while the water is run off similarly. After a period which may be approximately 30 or 60 minutes, all the vapor has been removed, the adsorbing power of the charcoal has been restored, and the adsorber is ready to function again, while adsorber No. 2 is steamed in turn. The life of the

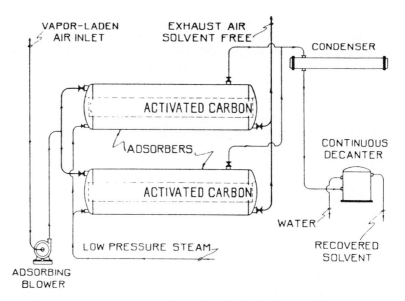

Figure 4.15 Flow diagram of one type of activated carbon solvent recovery plant. The two adsorbers used in rotation permit continuous operation.

carbon will depend on the type of carbon employed, on the solvents and impurities passed to it, and on the operating conditions.

The system may be modified to provide, in addition, a cooling and partial drying of the carbon bed after steaming, so that it is not placed in service again until cool and partially dried.

The opening and closing of the valves which divert the vapor-laden air from one adsorber to the other and the corresponding connection and disconnection to the exhaust line may be done manually or automatically. When the rate at which the air is fed in is constant, and likewise its vapor content, the system may be operated automatically on the basis of time. But even when the composition of the vapor-air mixture varies, automatic operation is available by using a vapor detector in the exhausted air. When the activated carbon in adsorber No. 1 is saturated, some vapor escapes, is detected, and the operation of an electrical device closes the inlet valve and opens or closes all other valves in the predetermined direction and order, so as to end the air mixture to adsorber No. 2. The system is then fully automatic.

Solvents which have been successfully recovered by the activated carbon adsorption method include methanol, ethanol, butanol, nine chlorinated hydrocarbons including perchlorethylene, which boils at 121°C (250°F), ethyl ether, isopropyl ether, the acetates up to amyl acetate, benzene, toluene, xylene, mineral spirits, naphtha, gasoline, acetone, methylethylketone, hexone, carbon disulfide, and others.

The volatile solvents recoverable by the activated carbon system or any other system are nearly all organic, and many of them form flammable or explosive mixtures with air. Such mixtures may lie between upper and lower explosive limits. The activated carbon system can avoid the explosive range by staying well below the lowest percentage of vapor which is still explosive; it functions well at very low concentrations. The system also recovers solvents efficiently even in the presence of water; the recovery efficiency is high (98 percent and 99 percent are not unusual); it may be fully automatic. The annual maintenance charge rarely exceeds 5 percent of the cost of equipment. The recovery expense may be as low as 0.2 cent per pound in some installations; it rarely exceeds 1 cent per pound.

Alternate Processes for Solvent Recovery

The recovery of solvent entrained in an air or gas stream may be accomplished by adsorption, condensation, or adsorption.

Adsorption

The selective removal of a constituent of a gas mixture by adsorption in a liquid in which only the constituent is soluble has found an application for the recovery of solvents. Although adsorption equipment is used for a wide range of applications, the recovery of solvents from very dilute gas or air mixtures of less than 1 percent is not economically achieved by the adsorption process.

Condensation

Recovery of solvents by direct condensation is normally used where pure solvent vapors, or solvent vapors at a high concentration, will be condensed by cooling. The method of cooling may be indirect by the use of a heat exchanger or by bringing the solvent vapors into direct contact with the cooling medium, such as water in a jet condenser. Examples of applications using condensation as a means of solvent recovery are dry-cleaning machines and many coating machines, solvent drying and printing operations, and so on.

GLOSSARY

Adsorber   A column filled with granular activated carbon whose primary function is the preferential adsorption of a particular type or types of molecules.

Adsorption   A reversible process in which fluid molecules are concentrated on a surface by chemical or physical forces, or both.

Bed Depth   The amount of carbon expressed in length units, which is parallel to the flow of the stream and through which the stream must pass.

Breakthrough   The appearance of an adsorbate of interest in the carbon-bed effluent at a predetermined concentration.

Breakthrough Curve   A curve which represents the concentration of adsorbate in the effluent stream as a function of time.

Countercurrent Efficiency   The unique advantage of a carbon column that permits partially spent activated carbon to adsorb impurities before the semiprocessed stream comes in contact with fresh carbon. This allows the maximum capacity of the activated carbon to be utilized.

Eductor   A device with no moving parts used to force an activated carbon water slurry through pipes to the desired location.

Makeup Carbon   Fresh granular activated carbon which must be added to a column system after a reactivation cycle or when deemed necessary to bring the total amount of carbon to specification.

Moving Bed   A unique application with granular carbons in which a single carbon column offers the efficiency of several columns in series. This is accomplished by the removal of spent carbon from one end of the carbon bed and the addition of fresh carbon at the other end with little or no interruption in the process.

Reactivation   The removal of adsorbates from spent granular activated carbon which will allow the carbon to be reused. This is also called *regeneration*.

Superficial Contact Time   The time required for a unit of liquid to pass through the bed. Superficial contact time is based on empty bed volume. The term *residence time* is sometimes used.

Carbon Adsorption Treatment of Hazardous Wastes

5

Commercial activated carbons appeared in the early part of the twentieth century, and their development was in applications in the sugar industry for decolorization and by the need to develop gas adsorbents to protect against the poisonous gases used in World War I. Carbon filters were developed in gas masks which resulted in a search for other applications. The use of carbons for decolorization and the purification of drinking water supplies has grown since that time.

The use of activated carbons for the treatment of liquid waste is now widely considered. The first large-scale system in the United States was the plant designed for tertiary treatment of 7.5 mgd of municipal wastewaters at South Lake Tahoe, California. Operations started in March 1968 and continue today with satisfactory results. Industrial wastewater treatment with activated carbon has accelerated with the advent of more stringent regulatory discharge requirements and the necessity to remove toxic compounds from effluents.

WASTES THAT CAN BE TREATED

In general, carbon adsorption is widely applicable to single-phase fluid waste streams, for example, liquid solutions or gas mixtures. Gas-phase and vapor-phase applications are discussed in Chapter 4. In actual applications, both aqueous and nonaqueous liquids are treated with carbon. The nonaqueous stream applications include petroleum fractions, syrups, animal and vegetable oils, and pharmaceutical preparations; color removal is the most common application in such cases. Waste treatment applications are limited to aqueous solutions.

Suspended solids in the influent, which lead to premature pressure carbon beds, should generally be less than 50 ppm to minimize backwash requirements. A downflow carbon bed can handle much higher levels (up to 2,000 ppm), but frequent backwashing is generally required. Backwashing more than two or three times a day is not desirable; at 50 ppm suspended solids, one backwash per day will often be sufficient. In any upflow packed bed excessive suspended solids can lead to clogging. Oil and grease should be less than about 10 ppm. A high level of dissolved inorganic material in the influent may cause problems with thermal carbon reactivation, such as scaling and serious loss of activity unless appropriate preventive steps are taken. Such steps might include pH control, softening, or the use of an acid wash on the carbon prior to reactivation.

Technically, there are no limits on the concentration of the solute(s) in the feed stream, but in practice about the highest concentration influent treated on a continuous basis is about 10,000 ppm TOC. An important factor concerning concentration to keep in mind is that the slopes of adsorption isotherms are almost always positive; the weight percent pickup (solute on carbon) will be higher for the more concentrated influent leading to more efficient solute removal. Additionally, if the carbon is to be regenerated after use, then higher concentration influents may also involve less frequent regeneration, especially if the slope of the adsorption isotherm is greater than one. When using carbon adsorption to clean up industrial waste streams, the controlling parameter is the quality of the effluent. Concentration of toxic materials in the feed must be balanced against the feed rate and the activity of the carbon to attain the effluent quality desired.

There are usually no limits on the flow rate of the feed stream. The main consideration is cost. Carbon adsorption benefits significantly from the economics of scale, especially the thermal reactivation portion. If carbon usage is less than approximately 1,000 lb/day, it may be cheaper to dispose of the carbon than to reactivate it thermally on site. Economics of carbon adsorption with on-site thermal reactivation will not be realized until carbon usage is above 8,000 lb/day. Using the lower figure of 1,000 lb/day and, as an example, for an industrial effluent (for example, effluent from a biological treatment system) that requires the use of 5 lb of carbon/1,000 gal of wastewater, the indicated flow is 200,000 gpd. This example should not be considered a cutoff point; in practice, industrial waste streams with flows in the range of 10,000–40,000 gpd have been successfully treated with carbon adsorption systems incorporating thermal reactivation. For lower flows, there are three options: (1) disposal by incineration or landfill of the carbon after use, (2) storage of the used carbon until an adequate amount is available to send to a central carbon reactivation service center, or (3) nonthermal regeneration with acid, base or solvent.

A wide variety of organic and inorganic solutes may be efficiently adsorbed on activated carbon. Applications involving organic solutes are more prevalent and will be most attractive when the solutes have a high molecular weight, low water solubility, low polarity, and low degree of ionization.

Highly soluble organics which often contain two or more hydrophilic groups are difficult to remove. For example, the adsorption of glycols from an industrial waste stream was not found to be feasible due to the low capacity of the carbon for the glycols. In another case, the treatment of wastewaters from a polyvinyl chloride production plant was found to be impractical; poor adsorption characteristics were attributed to the presence of long-chain organic soaps contained in the wastes. In the case of low adsorption or recovery of adsorbates, the higher costs involved in such processes may be offset by the value of recovered materials. Macromolecules, including certain dyes, may be too large to reach a significant fraction of the carbon's internal pores

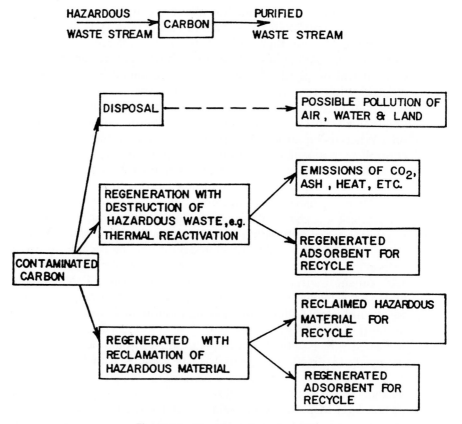

Figure 5.1 Steps in carbon adsorption.

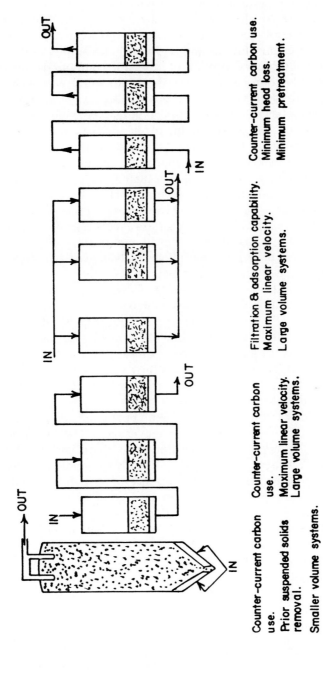

Counter-current carbon use.
Prior suspended solids removal.

Smaller volume systems.

Counter-current carbon use.
Maximum linear velocity.
Large volume systems.

Filtration & adsorption capability.
Maximum linear velocity.
Large volume systems.

Counter-current carbon use.
Minimum head loss.
Minimum pretreatment.

Figure 5.2   Adsorber configurations.

121

and, therefore, may be difficult to remove. Most industrial waste streams contain multiple impurities, some of which are easily adsorbed on carbon, while others are not. In considering the use of an activated carbon system, a series of laboratory tests is mandatory. Such tests should include both equilibrium adsorption isotherms and carbon column studies as discussed in earlier chapters. Carbon adsorption of inorganic compounds (for example, the removal of cyanide and chromium from electroplating wastes) has been found to be practical as well as a wide variety of other inorganics which will adsorb on activated carbon. However, adsorption may be quite variable from chemical to chemical, is likely to be highly pH dependent, and thermal or chemical regeneration may not be feasible. As a rule, strong electrolytes will not be adsorbed on carbon. Removal of inorganic solutes by carbon will generally involve influent concentration of less than 1,000 ppm, preferably less than 500 ppm. Processes other than physical or chemical adsorption may be involved. Plating may occur in some cases (for example, with ferric salts) and chemical reactions in others (for example, reduction of ammonia to chloramines followed by adsorption of the chloramines).

The process of carbon adsorption involves two basic steps, as shown in Figure 5.1. The waste stream is contacted with the carbon which selectively adsorbs the hazardous material and allows the purified stream to pass through. When the carbon reaches its maximum capacity for adsorption (or when the effluent is unacceptable for discharge, that is, when a breakthrough of hazardous material occurs), it must be removed for disposal, destruction, or regeneration. If a breakthrough of impurities occurs long before the full adsorptive capacity is reached, then either the rate of feed is reduced or the carbon is replaced. In some cases, carbon can be regenerated in such a way that adsorbed material is recovered.

Activated carbon is available in both powdered and granular form. Powdered carbon is less expensive and may have a slightly higher adsorption capacity but suffers from several drawbacks: (1) it is difficult to regenerate (without high losses); (2) it is more difficult to handle (the settling characteristics may be poor); (3) in the presence of suspended solids coagulation may occur (since powdered carbon is a good coagulant); and (4) larger amounts may be required than for granular systems to obtain good contact. Powdered systems are discussed further elsewhere in the text.

There are several ways that waste streams can be contacted with carbon, based on the choice of influent characteristics, effluent criteria, flow rate, and economics. The more common contacting methods, shown in Figure 5.2, are:

Method	Comments
Adsorbers in Parallel	• For high-volume applications • Can handle higher than average suspended solids (~65–70 ppm) if downflow • Relatively low capital costs

Method	Comments
	• Effluents from several columns blended, therefore less suitable where effluent limitations are low.
Adsorbers in Series	• Large-volume systems
	• Countercurrent carbon use
	• Effluent concentrations relatively low
	• Can handle higher than average suspended solids (~65–70 ppm) if downflow
	• Capital costs higher than for parallel systems
Moving Bed	• Countercurrent carbon use (most efficient use of carbon)
	• Suspended solids must be low (<10 ppm)
	• Best for smaller volume systems
	• Capital and operating costs relatively high
	• Can use such beds in parallel or series.
Upflow-expanded	• Countercurrent carbon use (if in series)
	• Can handle high suspended solids (they are allowed to pass through)
	• High flows in bed (~15 gpm/ft^2)

In the systems shown in Figure 5.2, flow rates will generally be in the range of 2–10 gpm/ft^2 for downflow operation, 2–7 gpm/ft^2 for upflow operation (without expansion) and around 15 gpm/ft^2 for upflow-expanded with about 10 percent bed expansion. Downflow operation may be either by gravity or pressure; the former is clearly less expensive but only applicable if suspended solids in the influent are low.

There is an additional method of contacting which is significantly different from other types of wastewater treatment system (for example, to the aeration basin) and yield significant improvements in pollutant removal. This method is currently being used in some industrial waste treatment plants and waste management facilities. Contacting by this method may be followed by biological regeneration of the carbon.

In some systems, contacting must be stopped periodically and the carbon column backwashed with water to remove suspended solids that have been filtered out of the wastewater by the carbon bed. In the case of downflow contactors, which may be designed to act in part as a suspended solids filter, frequent backwashing with bed expansions of up to 50 percent may be required. The backwash water is usually recycled to the primary sedimentation for treatment.

REGENERATION

If a large quantity of carbon is used, then regeneration is carried out to recover and reactivate the carbon. A common method is thermal reactivation which destroys the adsorbed organic solutes, but nondestructive chemical regeneration may be possible in some cases. Brief descriptions of some of the options follow.

Thermal Reactivation

Reactivation is carried out in a multihearth furnace or a rotary kiln at temperatures from 870–980°C as shown in Figure 5.3. Required carbon residence times in a furnace are of the order of 30 minutes. Steam is usually introduced to assist the reactivation at the rate of 1 lb steam/lb of carbon. Fuel requirements are about 4,000–5,000 Btu/lb of carbon. With proper control, the carbon may be returned to its original activity; carbon losses will be in the range of 4 percent to 9 percent and must be made up with fresh carbon. Afterburners (operating on the effluent vapors) ensure complete destruction of organic compounds, followed by a scrubber to remove any particulates. An additional 2,000–4,000 Btu/lb of carbon may be required to remove any particulates. An additional 2,000–4,000 Btu/lb of carbon may be required for operation of the afterburner.

Sometimes thermal reactivation of granular carbons may not be possible, for example, when (1) inorganic salts have deposited (or may deposit) on the carbon, though a preliminary acid wash may help avoid this problem; (2) the carbon contains adsorbates that would cause air pollution problems upon regeneration (NO_x, radioactive materials, and so on); (3) the adsorbates were explosive; or (4) the loaded carbon was excessively corrosive.

Alkaline regeneration for acid adsorbates

In some industrial wastewaters, the major organic component is acidic; adsorption carried out under acidic conditions, may be followed by removal of the solute (that is, regeneration) under basic conditions.

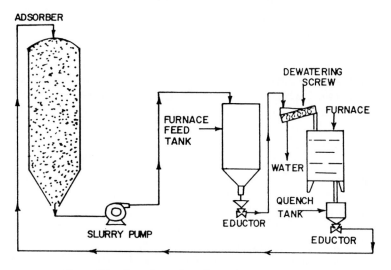

Figure 5.3   Granular carbon thermal reactivation cycle.

Acid regeneration for basic adsorbates

Certain basic adsorbates, adsorbed under basic conditions, may be recovered by washing with acid. This method has recently been found to be practical for the recovery of ethylene diamine.

Solvent regeneration

If the adsorbate easily dissolves in an organic solvent, it may be washed off the carbon and recovered after separation from solvent, for example, via distillation.

Steam regeneration

If the adsorbate is volatile enough, recovery may be carried out by passing steam through the carbon bed.

Biological regeneration

Attempts have been made to utilize biological regeneration. The effectiveness of this method will clearly depend on the biodegradability of the adsorbed organics. Complete regeneration is not achieved, and thermal regeneration may be required periodically. The preferred role of operation is to incorporate the carbon in a biological treatment system (for adsorption and assistance in biodegradation), recover the carbon with some activated sludge in the clarifier, and regenerate the carbon by aeration of the carbon/sludge mixture.

Output Emissions

Air and water emissions from carbon adsorption units employing carbon regeneration are relatively innocuous. The treated water is generally suitable for discharge to surface but may, in some cases, require reaeration to raise the level of dissolved oxygen, or other treatments to remove solutes not adsorbed by carbon. Other water streams associated with carbon systems such as backwash, carbon wash, and transport waters are recycled or sent to a settling basin. Emissions will result from thermal reactivation, but when afterburners and scrubbers are used, the emissions are nonpolluting. In some installations, particulates must be removed from this air stream via a cyclone and baghouse and this will result in a solid waste.

When carbon is not regenerated after use, then the spent carbon must be discarded. This will arise when (1) carbon usage is small, where generation is not economical, (2) the carbon was used to adsorb materials that are difficult to regenerate (for example, carbon loaded with explosives cannot be thermally reactivated), or (3) the carbon contained adsorbates that would cause serious pollution problems upon regeneration such as NO_x, radioactive materials.

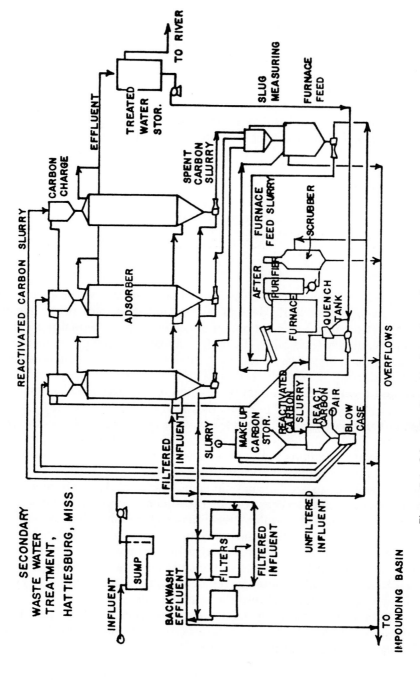

Figure 5.4 Schematic diagram of a carbon adsorption system incorporating thermal regeneration of the carbon.

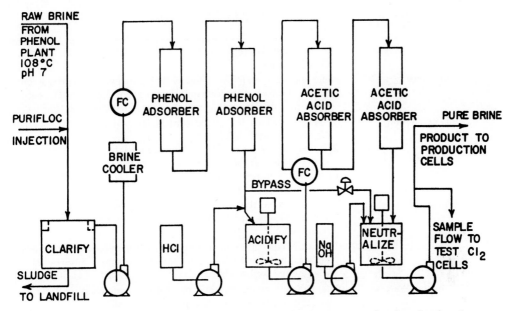

Figure 5.5 Schematic flow sheet of the adsorption process for phenol and acetic acid (to be followed by chemical regeneration and solute recovery).

Equipment and Materials

Figure 5.4 shows a complete system incorporating thermal reactivation, and Figure 5.5 shows an adsorption system used in conjunction with caustic regeneration. In the latter case, regeneration takes place in the same tank as adsorption and, thus, the only equipment items (of the total system) missing from Figure 5.5 are the caustic feed tank, the sodium phenate receiving tank, and associated pumps. Tanks must be constructed of stainless steel or coated (rubber or epoxy) steel to prevent corrosion. If gravity contacting is possible, then cement may be used for adsorber construction.

APPLICATIONS

The principal liquid-phase applications of activated carbon include (1) sugar decolorization, (2) municipal water purification (for water supply systems), (3) purification of fats, oils, foods, beverages, and pharmaceuticals, (4) industrial/municipal wastewater treatment, and (5) other liquid-phase applications for removal or recovery of dissolved materials.

TABLE 5.1   POTENTIAL APPLICATIONS FOR WASTE TREATMENT WITH ACTIVATED CARBON

Aqueous Wastes	With adsorbate concentrations up to about 10,000 ppm; SS 50 ppm, O & G 10 ppm
Organics	Prefer chemicals with low solubility, low polarity, low degree of ionization. Problem chemicals include acetone ethanol, glycol, soaps, and others that are of low molecular weight and/or high solubility.
Regeneration	(1) Thermal, which destroys adsorbates, is economical if carbon usage is above roughly 1,000 lb/day.
	(2) Chemical, may be used if one (or just a few) solute is present which can dissolve off the carbon. This allows material recovery.
	(3) Biological, if wastes are highly biodegradable. Virgin carbon activity not obtained after reactivation.
Disposal	Disposal of the carbon may be required if use is less than ~1,000 lb/day and/or a hazardous component mitigates against any form of regeneration.
Inorganics	Cr and CN currently removed in industrial applications. Other possibilities, listed in Table 4.1, are somewhat limited. Strong electrolytes are not well adsorbed.
Organic Wastes	Potential applications: Removal of color, oil, and grease, or other adsorbable impurities from solvents with resaleable potential.

Potential Applications to Waste Treatment

Activated carbon has a wide range of potential applications for waste treatment, with the largest number for the removal (and eventual destruction) of mixed organic matter from aqueous wastes. A summary of potential areas of applicability is shown in Table 5.1.

ENERGY, ENVIRONMENTAL IMPACTS

Energy requirements for carbon adsorption systems depend not only on the size of the system, but also on the method and frequency of regeneration required. When thermal reactivation is used, significant amounts of energy are used. Where reactivation is infrequent, total energy requirements may constitute less than 15 percent of total operating costs, and the electricity requirements associated with the adsorption process will account for most of these costs. For example, at the South Lake Tahoe unit, electricity requirements for the adsorption process accounted for about 11 percent of total operating costs, while gas and electricity requirements for the reactivation process amounted to only about 2 percent of total operating costs. This system was reactivating about 1.3 ton/day of carbon, and the energy requirements associated with both adsorption and reactivation included the use of 4,130 kWh/day of electricity and 113 therms of natural gas/day. Other estimates for energy requirements where thermal reactivation is used indicate that energy requirements are likely to be in the range of 10 percent to 25 percent of the total operating costs. Such requirements would be higher if very concentrated streams were treated. Fuel requirements for the thermal reactivation of carbon (natural gas, propane, LPG, or fuel oil) total around 6,000–8,000 Btu/lb of carbon, including afterburner fuel.

When thermal reactivation of the carbon is not employed, then energy costs may be less than 5 percent of total operating costs. If the carbon is not regenerated, then energy requirements may be even less. In fact, this used carbon may have fuel value and could be used in heating systems capable of burning solid materials.

There are no serious environmental impacts from carbon systems using thermal reactivation since the only output streams are (1) a purified liquid or water and (2) furnace gases, which have been through an afterburner and a scrubber and, in some cases, a dust filter. When carbon is chemically generated, the major output streams consist of (1) the purified liquid, (2) a recovered material, and (3) used carbon, since carbon will lose some activity in successive chemical regenerations and will have to be replaced periodically. In systems where the carbon is not regenerated at all, impacts result from the disposal of the carbon and possibly from contained toxic or hazardous materials.

Economic Factors

Variables or alternatives in the design and operation of a carbon treatment unit can affect the economics and include:

- type of carbon—granular or powdered
- contact time
- flow rate
- configuration—series, parallel, or moving bed
- number of stages
- flow direction—packed or expanded, upflow or downflow
- hydraulic force—pumped or gravity
- carbon capacity
- method of regeneration—thermal, chemical, or none

Different carbon adsorption systems may result in unit costs that will vary by over two orders of magnitude.

Capital and operating costs for carbon systems employing thermal regeneration are a function primarily of: (1) the carbon exhaustion rate, usually expressed as pounds of carbon reactivated per volume of liquid treated; and (2) superficial liquid retention time, that is, the time that the liquid would take to fill the volume of the carbon bed. This is a direct function of the liquid flow rate and the carbon volume.

The total capital cost is established by the volume of the carbon beds and the size of the reactivation furnace. The operating costs are determined by the carbon exhaustion rate since the largest variable is usually the cost of the makeup carbon. These two variables are related for a specific system.

Capital cost estimates will depend on the optimum superficial retention time after the exhaustion rate has been determined, the adsorption and reactivated portions of the system may be sized. Capital cost versus size curves can be computed for any given system.

Costs for thermally reactivating carbon increase rapidly as the volume regenerated falls below 1,000 lb/day, and at some point regeneration will be more expensive than replacement. For high-volume systems (for example, more than 10,000 lb of carbon regenerated per day), regeneration costs will not decrease substantially for larger systems. Capital costs for a thermal reactivation system in an adsorption system treating relatively concentrated wastes (for example ~1,000 ppm adsorbate concentration) can amount to 45 percent to 50 percent of total capital costs. This percentage will increase with the adsorbate concentration increases. An alternative to in-house capital investment has been offering services which involve the leasing of adsorption equipment and the regeneration of the carbon at a regional regeneration center. Applications involve treatment of streams as small as 7–10 gpm and as high as several mgd. Almost any system configuration can be put together with the prefabricated units that are used for the adsorption system. With

Figure 5.6   Skid-mounted portable carbon adsorption unit.

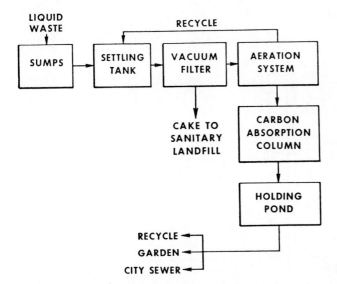

Figure 5.7 Pilot adsorption system.

this service, a company can avoid all capital costs (except some site preparation costs) and instead pays a service fee. Such systems benefit from economies associated with (1) the use of predesigned (and partially prefabricated) units, (2) quantity purchases of carbon, and (3) volume regeneration. These are offset by the costs of trucking the carbon between the customer and the center, and by fees charged. Figure 5.6 shows a skid-mounted portable carbon adsorption unit. Figure 5.7 is a schematic of a pilot adsorption system. Figure 5.8 shows pretreatment filtration for an adsorption system and Figure 5.9 illustrates a carbon adsorption/filtration plant for pesticide waste treatment.

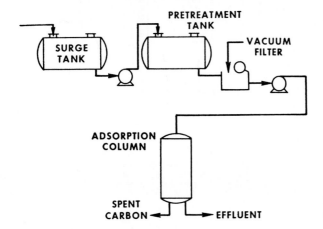

Figure 5.8 Pretreatment filtration adsorption system.

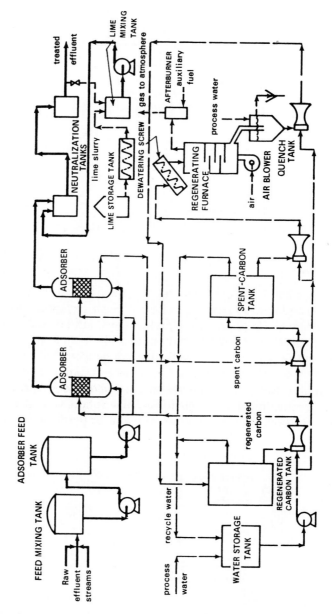

Figure 5.9 Carbon adsorption/filtration plant for treating pesticidal wastes.

Advantages for Carbon Treatment of Toxics

Major benefits of carbon treatment include:

- High organics removal.
- Applicability to a wide variety of organics.
- Capability for inorganics removal such as CN, Cr, and others.
- Possible material recovery of organics and inorganics in a few cases (for example, phenols, acetic acid, ethylene diamine, Cr, and others).
- High flexibility (for example, rapid start-up and shutdown).
- Low sensitivity to feed variations (flow rate or concentration).
- Insensitivity to toxic materials.
- Minimum land area.
- Destruction of wastes when using thermal regeneration.

Limitations of carbon treatment include:

- Relatively high capital and operating costs, especially when thermal reactivation is used.
- Limited, generally, to wastes with less than about 1 percent organic content, or less than about 5 percent if adsorbates are to be recovered.
- Cost savings associated with thermal reactivation may only be realized if carbon usage is above about 1,000 lb/day or if a centralized reactivation facility is available.
- If carbon cannot be regenerated thermally or chemically, then it must be disposed of.
- Low tolerance of suspended solids—should be less than about 50 ppm to reduce the need for frequent backwashing.
- Inability to remove low molecular weight and/or highly soluble organic chemicals (for example, methanol, ethanol, glycol, soaps, and others).
- Problems with thermal reactivation systems.

Certain process modifications may improve the applicability of the carbon process. One is the upflow, expanded-bed flow configuration which has the capability of handling much higher levels of suspended solids. Another is the use of powdered carbon which is cheaper than granular carbon and may have higher adsorption capacity. However, powdered carbon is not economical to regeneration because of high losses (around 15 percent).

Carbon adsorption may be considered a viable and economical process for organic waste streams containing up to 1 percent to 5 percent of refractory or toxic organics; its applicability for inorganics removal, though demonstrated in a few cases, is probably much more limited.

Carbon adsorption systems have been demonstrated to be practical and economical: (1) for the reduction of COD, BOD, and related parameters in secondary municipal and industrial wastewaters (*tertiary treatment*); (2) for the reduction of COD, BOD, and related parameters in primary municipal and industrial wastewaters (*physical-chemical treatment*); (3) for the removal of toxic or refractory organics from isolated industrial wastewaters; (4) for the removal and recovery of certain organics from wastewaters; and (5) for the removal, at times with recovery, of select inorganic chemicals from aqueous wastes.

The technology for systems employing adsorption of organics followed by the thermal reactivation of the carbon (which destroys the adsorbates) is relatively well developed. Systems employing chemical regeneration, with recovery of adsorbates, are less developed; the potential for carbon systems in such applications (that is, for the recovery of both organic and inorganic chemicals) is greater than is generally realized.

Carbon adsorption systems, especially those employing thermal regeneration, are generally considered to involve both high capital and operating costs. The process economies achievable with thermal reactivation of the carbon (on site) may only be realized if carbon usage is above 1,000 lb/day; if wastewater flows are above 10–50,000 gpd, and the adsorbate concentrations are not too low, thermal reactivation will probably be economical. Below these cut-off points it is more economical to purchase new carbon or to use the services of a centralized reactivation facility.

Carbon adsorption can be given serious consideration when it is desirable to remove mixed organics from wastewaters, to remove select inorganics from wastewaters, or to recover select organic in inorganic species from aqueous solution. The concentration of adsorbates in the influent should be less than 1 percent when recovery is not involved; concentrations up to 5 percent are acceptable when recovery is feasible and justified.

PESTICIDE WASTE TREATMENT

The use of pesticides has become extremely important. Benefits obtained from their use include increased production of food and fiber and increased freedom from disease and obnoxious plant and animal life. However, these benefits have not been gained without some undesirable side effects, such as direct effects on nontarget organisms, the indirect unbalancing of delicate ecosystems, and the environmental contamination by persistent pesticides, which may tend to be biologically accumulated in food chains. In addition, the possible long-term effects of low levels of pesticides on humans are causes for serious concern.

The production and use of pesticides are not new or even of recent origin. People have investigated the minerals and the plant and animal life around them for their value as medicinals, in the production of food, in warding off attacks by obnoxious or dangerous insects, and in warfare. However, tremendous growth has occurred during the recent past in the number of pesticides available, the variety of applications, and the volume of active ingredients produced and their formulated products. A broad definition of *pesticides* includes: rodenticides, insecticides, larvacides, miticides (acaricides), molluscicides, nematocides, repellants, synergists, fumigants, fungicides, algicides, herbicides, defoliants, desiccants, plant growth regulators, and sterilants. On this basis, more than 1,500 chemical pesticides have been produced. About 275 pesticides are of commercial importance and perhaps as many as 8,000 individual formulated products are prepared for specific uses and methods of application.

Special categories include petroleum oils (of which some synthetic and refined products are used directly as insecticidal sprays, but most of which are used more extensively as diluents and carriers and as wood preservatives), creosote and coal tar (general wood preservatives), aromatic solvents, and the dry carriers and diluents. These materials have been of little environmental concern compared with the more active ingredients in most pesticides.

The technology for the production of pesticides varies widely depending on the properties of the compounds and the position of the company in terms of raw materials, patent position, and sales structure. Production of the early inorganic pesticides was essentially by batch processes and since most of the ingredients were nonvolatile solids or handled in aqueous media, precautions to avoid occupational exposure or environmental pollution were probably minimal. The rise of the petrochemical industry brought an increasing capability to conduct continuous process manufacturing; much of this technology was tapped by the synthetic organic pesticide industry, and semi-continuous operations are common. In fact, a number of the petrochemical companies had the raw materials and technical know-how to move rapidly into the pesticides field.

The production of synthetic organic pesticides in general, however, involved the use and handling of more hazardous materials (for example, chloral and chlorobenzene for DDT) than did the inorganics and, in many cases, the production systems were more carefully controlled or enclosed to avoid occupational exposure. The advent of the chlorinesterase-inhibiting, organophosphate, and carbamate insecticides lead, of necessity, to almost completely enclosed or controlled production systems. Many of the highly effective or specific recently developed materials require considerably more sophisticated production chemistry and technology; thus, the product is quite expensive and more care is exercised to recover the product from waste streams.

Major pesticide producers have, on the whole, extensive wastewater treatment facilities. Many of these are newly modified. The disposal of liquid wastes from pesticide manufacture varies widely with different companies, products, and geographical locations. Methods being used include many varieties of neutralization, oxidation, settling and holding ponds, and also secondary and biological waste treatment plants (all of which are followed by discharge to a stream or lake); evaporation basins (which have no out-fall); deep well disposal; deep ocean disposal; and incineration.

Historically production processes as employed for some product lines led to sizeable losses of active ingredients, toxic raw materials, by-products, and so on, and these were often not detoxified by the waste treatment facilities. For example, discharges of active ingredients range from a few pounds to over 1,000 lb/day for some products.

Nearly all of the basic facilities and equipment in use for pesticide manufacture and formulation were designed and built prior to the present age of intense concern about environmental quality. This situation is not unique to the pesticide industry but has prevailed with most manufacturing facilities and processes that may have been in use. However, this problem is of special importance in the pesticide industry because of the biologically active chemicals that are apt to have higher potential for causing environmental damage than do the effluents discharged from many other manufacturing processes. Activated carbon technology comes into focus as a viable solution for the treatment of pesticidal waste streams.

By-Products and Wastes

Virtually every pesticide production process produces aqueous or gaseous streams and frequently solid wastes which contain unreacted ingredients, unrecovered products and solvents, and unavoidable or undesired by-products. Extensive efforts are usually made to minimize by-products and to recover, recycle, or otherwise prevent these process losses from occurring. For each process, however, a balance point is eventually reached between the expense of recovery and the value of the recovered product. Economic considerations are frequently dominated, and process losses are included as unavoidable costs. With the emphasis on environmental contamination, efforts have been made to recover many previously lost materials—even when economics indicated that it is more expensive to do so. Most pesticide manufacturers have invested in or built extensive waste treatment facilities wherein those wastes that cannot be recovered are degraded to acceptable levels or disposed of by state-approved methods. A summary of principal wastes generated by producers of the key pesticides is shown in Table 5.2. The importance in removing some of these compounds is illustrated by their toxicology, as demonstrated in Figure 5.10.

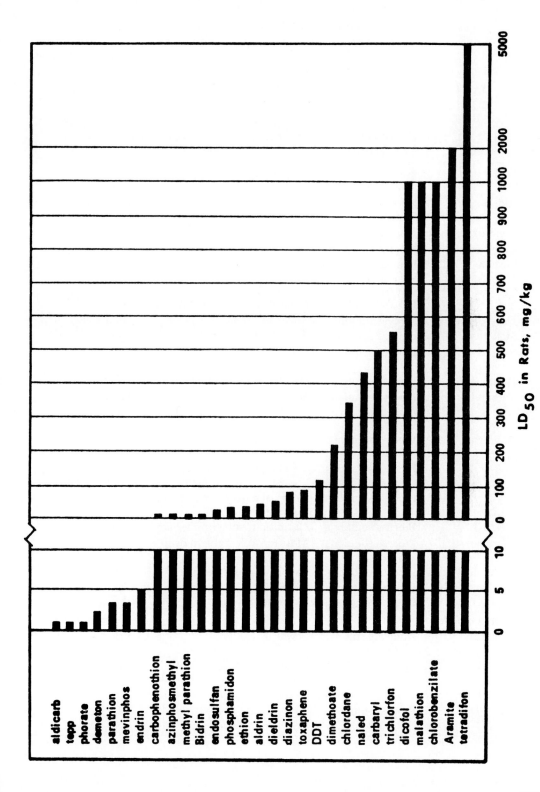

Figure 5.10   Acute oral toxicity.

TABLE 5.2   SUMMARY OF MANUFACTURING WASTES AND DISPOSAL

| Pesticide | Liquid wastes | | Solid or other wastes |
	Source	Disposal	Source
DDT	Processing solutions	Evaporative basin	Reactor solutions
Aldrin	Floor washings, etc.	Evaporative basin	Lime slurry
Dieldrin	Process solutions	Evaporation basin	Filter solids
Chlordane	Process solutions	Deep well	Filter solids
Toxaphene	Pinene-camphene plant		
	Process solutions	Neutralize, waste treatment	Filter solids
Disulfoton	Process solutions	Secondary treatment plant	Filter solids, etc.
Malathion	Process solutions	Barge to deep sea	Filter solids
Phorate	Process solutions	Barge to deep sea	Filter solids Mercaptan losses
Parathions	Process solutions	Waste treatment plant	H_2S, S
Carbaryl	Process solutions	Secondary waste treatment	H_2, $COCl_2$, amine Heavy residues
Aldicarb	Process solutions	Neutralize, secondary waste treatment	Process vents
2, 4-D	Process solutions	Trickling filter, biological waste treatment plant	Filter solids and still bottoms
2, 4-D	Process solutions	Charcoal adsorption/filtration treatment	
2, 4, 5-T	Process solutions	Chemical treatment, discharge	Solids
Atrazine	Process solutions	Most to river; some to deep well	
Trifluralin	Process solutions	Biological waste treatment	NO_x
Alachlor	Process solutions	Discharge	Solvent
Captan	Process solutions	Hold, discharge	Gas streams
Methyl bromide			Gaseous wastes
Pyrethrin	Aqueous still bottoms	Sewer	Process solids Filter solids
Bacillus t.	Process solutions	Sterilized, biological waste treatment	Process air
Bacillus t.	Process solutions	Evaporation pond	
$HgCl_2$-$HgCl_2$	Process solutions	Hg-recovery; treated; liquid discharged	NO_x H_2S

Adsorption of Pesticides

Many adsorption studies have been conducted on a large number of pesticides. The following are a list of adsorbents used and a list of pesticides and some results of the studies.

Adsorbents: Activated carbon; saturated clay systems (H/Al, Ca Mg, K); humic acid, New Jersey soils, muck soil (organic soil); bentonite clay; cation exchange resin; anion exchange resin; EPK (Edgar Plastic Kaolin from the EPK Company); hydrousaluminum silicate; USP (Fisher Scientific Company); and Diluex (hydrous magnesium aluminum silicate).

Pesticides: DDT, DDD; lindane; parathion; dieldrin; endrin; aldrin; chlordane; malathion; captan; BHC; 2,4-D derivatives; 2,4-DCP; 2,4,5-T ester; toxaphene; rotenone; CIPC; DCPA; Ij-DNBP; trifluralin; diphenamide; amiben; paraquat; and linuron.

Results of activated carbon removal:

- Activated carbon on herbicides and pesticides has shown that it is successful in reducing the concentration of these toxic compounds. Included are such widely used herbicides and insecticides as BHC, DDT, 2,4-D, toxaphene, dieldrin, aldrin, chlordane, malathion, and parathion.
- Activated carbon on the removal of fish poisons has shown that rotenone, toxaphene, and the solvents and emulsifiers present in all commercial fish poison formulations are removed. This broad activity also eliminates odors.

TABLE 5.3   EFFECT OF ACTIVATED CARBON ON HERBICIDES AND INSECTICIDES

Pesticide	Concentration treated with carbon (ppm)	Carbon dosage used (ppm)	Threshold odor values Before	Threshold odor values After	Threshold odor units removed per ppm carbon	Concentration after calculated odor reduction (units)	Treatment determined by chemical analysis (ppm)
Parathion	10	10	50	4	4.6	0.8	2.6
BHC-37	25	5	70	6	12.6	0.22	0.08
Malathion, 50%	2	10	50	4	4.6	0.08	0.25
2,4-D, 23.5%	6	20	50	3	2.4	0.085	1.38
2,4-D, 11.7%	1	10	70	2	6.8	0.005	L[a]
Chlordane, 6%	50	10	50	1.4	4.9	0.084	L[a]
DDT, 50%	5	2	70	4	33	0.15	L[a]

[a] Concentrations were too low to determine by available test methods.

- Removal of malathion, 2,4-D, aldrin, dieldrin and DDT (by means of adsorption on activated carbon) shows that substantially 100 percent removal could be obtained for all levels of pesticide used.
- Activated carbon adsorption is measured by the reduction in biological activity as follows: CIPA, trifluralin, 2,4-D, diphenamide, DCPA DNBP, and amiben. CIPC and trifluralin, which were the most readily adsorbed by activated carbon, were desorbed the least. 2,4-D was readily desorbed from both activated carbon and bentonite clay.
- Activated carbon easily removes DDT, lindane, parathion, dieldrin, endrin, and 2,4,5-T ester (for initial concentrations of 10 ppb). Table 5.3 shows activated carbon treatment effects.
- High efficiency of adsorption for BHC, chlordane, DDT, aldrin, DDD, and endrin has been demonstrated.

COMBINED POWDERED ACTIVATED CARBON-BIOLOGICAL TREATMENT

A recently developed application of activated carbon to wastewater treatment involves the addition of powdered activated carbon to the aerator of a submerged culture biological process. This combined powdered activated carbon-biological process is often referred to as the PACT Process (*Du Pont Service Mark). Applications have covered carbon addition to activated sludge, aerated lagoons, contact stabilization-type processes, and even rotating disk contactors. The most frequently mentioned application of powdered carbon to biological systems is to the activated sludge process (Figure 5.11).

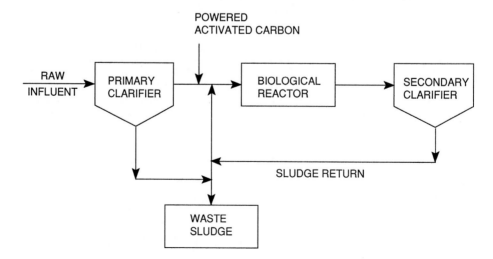

Figure 5.11 Application of activated carbon to an activated sludge process.

APPLICATIONS

Technical advantages achieved by adding powdered activated carbon to a submerged culture biological process have included:

- Added system stability against shock loading, temperature changes, and so on.
- Improved nonbiodegradable organics removal.
- Color removal.
- Improved removal of compounds on EPA's priority pollutant list.
- Resistance to biologically toxic substances in the wastewater.
- Improved hydraulic capacity of existing plants.
- Improved nitrification of ammonia.
- Suppressed foaming in aerators.
- Improved sludge settling/thickening/dewatering.
- Reduced sludge bulking.

Added System Stability

For any number of reasons, activated sludge processes can become unstable, resulting in deterioration of effluent quality. System instability is very desirable since wastewater treatment plants are expected to operate 100 percent of the time to achieve effluent quality goals, often set by regulatory requirements regardless of the quality or quantity of feed. Carbon addition increases average BOD removals from 72 percent to 89 percent, and also significantly decreases the variability in effluent quality (Figure 5.12).

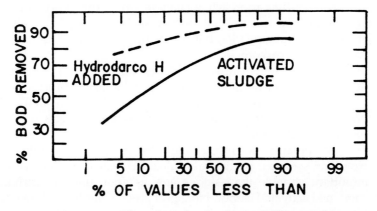

Figure 5.12 Effect of powdered carbon on BOD removal.

Improved Nonbiodegradable Organics Removal

The addition of powdered activated carbon to a biological system can be expected to be of most benefit in removing carbon-adsorbable materials. As a general rule, activated carbon does not easily adsorb the readily biodegradable compounds like methanol and acetic acid but is very good for adsorbing relatively nonbiodegradable molecules like substituted aromatics.

PROCESS VARIABLES

Process variables found to be significant for the combined activated-carbon biological process include all of those normally associated with biological processes plus those associated with the addition of the powdered carbon.

The most significant new variable introduced by the carbon addition to biological treatment is carbon dosage. This is also greatly affected by carbon type. For the organic chemical industry waste, both the carbon dose and carbon type are very significant in predicting effluent quality.

Carbon concentration in the aerator is the important variable. For steady-state operations, the carbon dosage and aerator carbon concentration are related by:

$$MLSS_c = \frac{24\theta_c(X_c)}{\theta}$$

where

$MLSS_c$ = aerator-mixed liquor suspended solids due to carbon, mg/1
θ_c = sludge age, or sludge residence time, days
X_c = carbon dosage, ppm
θ = aeration time, hr

Still another approach is taken in which the performance is plotted not against carbon dosage, but carbon surface area. For a given carbon this amounts to the same plot, but using surface area appears to allow correlation with different carbons.

Many designers now use sludge age, or sludge residence time, as a design parameter for submerged culture biological processes.

There are some indications that a combined activated carbon-biological process performs better on some wastewaters at longer residence times and the same sludge age.

Temperature is perhaps the most important variable in the design of biological systems, particularly in cold climates. In some cases, it is a more important variable than sludge age, aeration time, or carbon dosage. The combined activated carbon-biological process has been shown to be extremely effective for low-temperature operation.

Once the process variables are established, any combined activated carbon-biological process must be transformed into a practical process in reliable hardware. Perhaps the first consideration of a combined activated carbon-biological process is the choice of aeration equipment. Experience suggested that powdered carbon addition does not limit this choice much, if any. Carbon addition to biological systems has been successfully demonstrated in many types of aeration equipment.

The addition of carbon has been shown to greatly facilitate handling a biological sludge. The final step for some plants might be carbon regeneration, if economics on carbon dose and regeneration furnace investment suggest the practicality. This can be done in any number of ways.

6

Carbon Regeneration

Activated carbon has been used as a sorbent for many years and regeneration of the spent carbon has been practical for about the same time. Aspects of activation, sorption, and regeneration as well as general applications have been discussed in the preceding chapters. In recent years, activated carbon has received considerable attention as a means for preventing the emission of pollutants into the environment by removing pollutants from both liquid and gaseous waste streams.

The removal of pollutants from waste streams will add to the cost of an industrial operation or a municipal sewage treatment plant. For this reason, the development of technology for producing less expensive types of activated carbon and for regenerating the spent carbon make the use of activated carbon viable as a pollution control measure.

Carbon systems usually consist of two distinct operations: (1) the contact (adsorption) process, and (2) a carbon regeneration system. A typical arrangement of equipment for use in a granular carbon system is shown in Figure 6.1. Process liquor is pumped in the column, which is packed with granular carbon and the purified process liquor is removed at the top of the column. Spent carbon is removed from the column periodically for regeneration and then fed back into it. Fresh carbon for makeup is added to the top of the column.

The operation of the carbon regeneration furnace can best be described by considering it as part of a regeneration system. All the equipment is directly

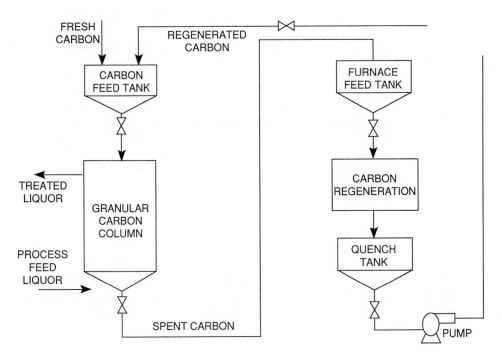

Figure 6.1 Typical equipment arrangement for granular activated carbon.

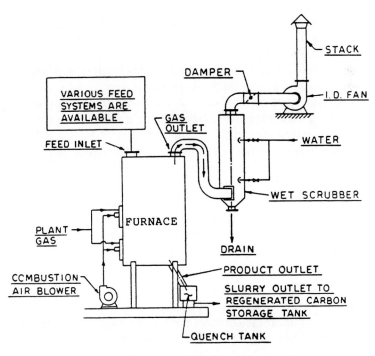

Figure 6.2 Carbon regeneration furnace.

related to the furnace (as shown in Figure 6.2). The regeneration system consists of the following operations:

- Conveying of the slurry.
- Receiving, dewatering, and feed tanks.
- Feeding.
- Regeneration furnace (MHF = Multihearth furnace).
- Quench tank and conveying.
- Off-gas equipment.
- Controls and instrumentation.

As the carbon is discharged from the carbon columns, it has to be conveyed to a dewatering device before feeding it to the furnace. The carbon is conveyed in a slurry form, which is handled by the following equipment:

- *Centrifugal pumps:* Normally, the pumps used for this service are rubber lined, not only to protect the pump material, but also to minimize degradation of the carbon particles. There are several types and makes that are applicable to this application.
- *Diaphragm pumps:* As the name implies, these pumps are actuated by means of a diaphragm. There is one drawback with these units: They require at least a 3-ft suction head, which makes it difficult to fit in some installations.
- *Eductors:* Another accepted method is the use of stainless steel eductors. An auxiliary source of motive power is required, such as a 60–80 lb water supply or, better still, a centrifugal pump to supply water at the required pressure. Hard chrome plating is also used in some eductors to reduce the price.
- *Blow cases:* This is a chamber which is pressurized with air to propel the slurry forward. Automatic valves are used to cut the feed to the chamber, close the atmospheric relief, open the air and the discharge. They are rather expensive and not many are in use now.
- *Torque flow pumps:* This is a special type of centrifugal pump with a wide impeller shaped with sweeping curves that effectively minimize carbon particle degradation.

The carbon slurry is difficult to convey and care should therefore be taken in the design of the piping.

An important consideration is to keep the slurry moving, which will keep the carbon in suspension. To accomplish this, the velocity in the pipes should be above 2.5 ft/sec and preferably above 3.5. Also, elbows should not be used. In their place, long radius bends with a radius equal to at least three times the pipe diameter should be used. Flushing or backwashing connections at certain locations such as at the foot of the risers and other key places have to be considered when laying out the piping. In addition, sight-flow indicators can be installed closer to the pumps or other accessible location for a visual check that the slurry has not settled and plugged the piping. All valving should be constructed with stainless steel ball valves. If tight shut-off is not required, a butterfly valve may be used.

The piping should be kept as short as possible; that is, the regeneration equipment should be placed as close as possible to the carbon columns.

Dewatering and Feed Tanks

The carbon slurry has to be received and dewatered before feeding it to the furnace. There are two basic carbon column operating systems: the batch and the intermittent or slug type. Depending on which system is used, the receiving, dewatering, and feed operations are performed differently.

In the batch system, several columns or cisterns operate in series, and one extra column is on standby. When a column is operating and then is switched to standby, all the carbon in that column has to be regenerated within a predetermined length of time. The complete column, therefore, has to be transferred to the receiving and/or dewatering tank and then fed to the furnace. As the columns are spent and a new column put in service, it is placed last in the sequence to have the best carbon in contact with the cleanest water.

Dewatering and feeding may be accomplished by one of the following methods:

1. On large installations, such as municipal wastewater or sewage water treatment, a single floor-mounted tank, the same capacity as the carbon columns, may be used. It functions as a storage tank to have the carbon column empty to receive the regenerated carbon charge. Another similar tank could also be used as storage for the regenerated carbon. The tank is provided with a shallow cone bottom, an overflow screen, flushing nozzles, and a discharge nozzle. Preferably, it should be covered (Figure 6.3). The carbon slurry from this tank is discharged into a dewatering screw conveyor which is described later.
2. On small installations, which are more frequently encountered in industry, or test units such as at Pomona, an elevated tank above the furnace would be used (Figure 6.4). The savings from eliminating the pumps, valving, and slurry lines are offset by the additional cost of a structure to support the tank and the rotary feed valve (described later). The tank is fitted with a screened overflow. A dewatering screw conveyor is used for feeding the furnace.

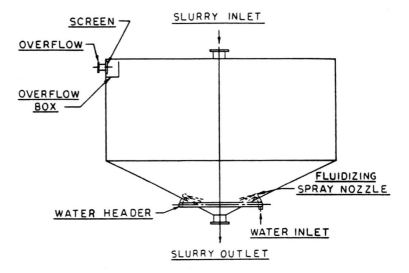

Figure 6.3 Dewatering the storage tank.

3. Another method is to use a pyramided tank at grade level with an inclined dewatering screw feeder, as shown in Figure 6.5. Tank materials are carbon steel, which is used frequently, stainless steel, stainless clad, resin, or epoxy-lined are used where iron contamination in the product is undesirable, or corrosion by the adsorbate in the carbon is possible.

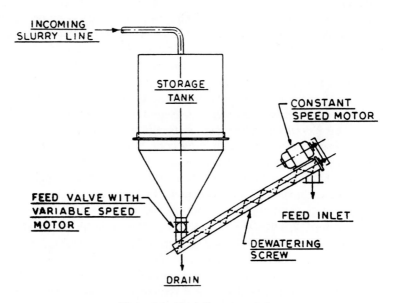

Figure 6.4 Typical storage tank.

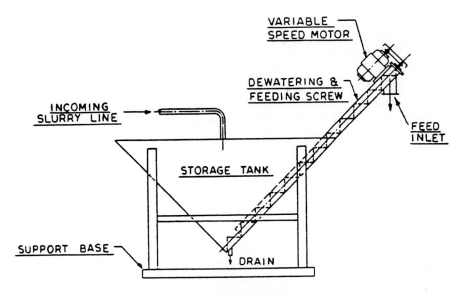

Figure 6.5   Storage tank.

Intermittent systems are also known as the slug or moving-bed type. A slug of spent carbon is intermittently withdrawn from the carbon column. Flow through the column is only stopped briefly during the time necessary to take the slug out and replenish it with regenerated carbon at the top. Carbon in the column moves countercurrent to the fluid being processed.

Figure 6.4 shows the method used for receiving, dewatering, and storage—a two-tank arrangement. The top tank has a special type of screen fitted to the coned bottom; the water is drained into a special box and out. These tanks have to be fitted with vibrators and also with air backflushing to dislodge the dewatered carbon and discharge it. The screen is a special nonclogging type and is used in carbon installations. It is mounted on a specially made underdrain frame, which is stitch-welded to the tank bottom. This screen is always fabricated of stainless steel. Capacity of the tank should be enough to take in one slug from the carbon column. The slug size is established by experience in the behavior of the carbon in that particular application, which determines the size and frequency of the slugging operation.

Underneath the dewatering tank, there is a discharge valve. Normally, this valve is pneumatically operated by remote control from the panel. The control circuit is interlocked with the slurry pump starter so that if the valve is open, the pump cannot be started. In this way, the water from the slurry cannot be accurately dumped into the hot furnace, thereby causing extensive damage to the brickwork. The valve can be with a special ring seat or a positive, closing-type valve. When the valve is opened, the dewatered carbon is dropped from the top tank into the bottom tank, which is called a feed or surge tank. The latter is usually slightly larger in size to ensure that all the carbon can be dropped with no overflow.

The materials of construction are carbon steel, stainless steel, stainless steel clad, epoxy lined, and/or resin. The surge tank has a flanged discharge where the screw feeder is attached.

Two other methods are used for receiving, dewatering, and storage. These are shown in Figures 6.4 and 6.5. The tank sizes for an intermittent system are smaller since they only have to receive a slug which may be no more than 100–200 ft^3 in size.

Feeding Equipment

Feeding equipment can be divided into belt, screw feeder, and rotary valve types.

The *belt conveyor* can be used for feeding but must be wide to fit the discharge. In turn, the tank discharge cannot be reduced beyond certain limits, otherwise bridging will occur. Due to its width, it has to be run extremely slowly as the carbon feed rates are normally very small. Therefore, this requires an intermittent operation to fit the low rates. In addition, a separate seal, usually a rotary air lock, has to be provided at the furnace inlet. Furthermore, wet carbon may stick to the belt, which will result in a dirty operation. The one advantage of the belt is that carbon particles do not undergo deterioration.

Screw conveyors can be used for feeding. Several advantages can be realized with this type of equipment. It seals the furnace inlet because by using a covered trough and a flanged connection to the feed tank, the carbon itself serves as the sealing medium, and it eliminates the rotary air lock. It can be slowed down sufficiently to feed the very low feed rates to the furnace and the drive can be of the variable-speed type so the feed rates can be varied at will, and it makes a clean installation. The screw feeder is normally furnished in stainless steel construction, although in some cases, a carbon steel trough can be used.

The screw conveyors that can be used on installations include:

- A plain screw feeder (Figure 6.6).
- A dewatering inclined conveyor, in which case it is fitted with a dewatering screen at the back end (Figure 6.4).
- A dewatering and feeding conveyor (Figure 6.5).
- Another version of the dewatering screw that has been used successfully in a number of installations (Figure 6.7). It consists of an inclined screw with a box-like back end and an internal overflow weir. The drive is usually a variable-speed motor reducer.

Rotary air lock fitted with a variable-speed drive is typically used as shown in Figure 6.4 in connection with the plain receiving tank and an in-

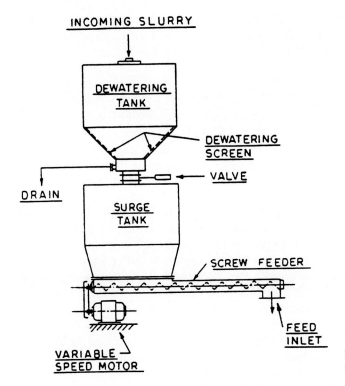

Figure 6.6 Dewatering the surge tanks.

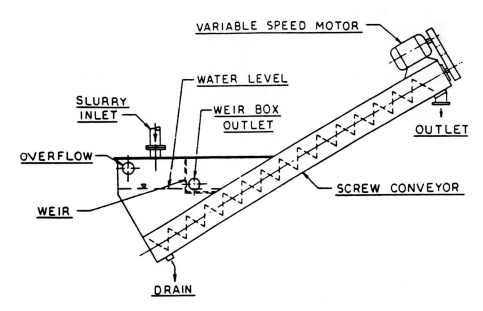

Figure 6.7 Inclined screw feeder.

clined screw conveyor. Materials of construction can be chrome-plated cast iron or nickel-hard casing. The rotary valve is made of hard-faced tips or stainless steel replaceable blades.

Multihearth Furnace

The typical multihearth furnace employs a simple design approach. It consists of a steel sheet lined with refractory inside. This refractory can be a castable as used in the 30-inch units, or brick as used in the larger sizes. The latter can also have 4.5 inches of insulating blocks which make the walls a total of 9 inches thick for the small furnaces or 13.5 inches on the larger furnaces where high temperatures are used.

The interior space of the furnace is divided by horizontal brick arches into separate compartments called hearths. Alternate hearths have holes at the periphery or at the center for the carbon to drop through from one hearth to the next. Through the center of the furnace goes a rotating shaft driven at the bottom by a speed reducer with variable-speed drive. It is scaled at the top and bottom by special sand seals to prevent air or gas leakage. The shaft is hollow and has sockets where arms, called *rabble arms*, are fitted. An inner tube in each arm and in the shaft provides the means for air cooling of both to prevent damage by the intense heat. This cooling air is blown in through a special connection at the bottom of the shaft. The arms are, in turn, fitted with rabble teeth, placed at an angle, and impart a motion to the carbon when the shaft is rotated, moving it in some hearths and out in the others. The hearths that move the material out and which have the peripheral holes, called *drop holes*, are fitted with lute caps. A lute cap is just a disk attached to the shaft which prevents the material dropping from the hearth above. Burners are attached at certain locations in the shell to heat the furnace to the required temperature. These burners can be of the nozzle-mixing or premixed type and are normally set so excess air can also be introduced through them for the oxidation of the organic impurities in the carbon. The burner system includes the usual complement of accessories such as spark-ignited pilots, solenoid valves, mixers, regulators, safety shut-off valves, combustion air blowers, and so on. Fuel used can either be No. 2 diesel oil or gas, either natural, propane, or butane. The shell is also provided with nozzles for injecting steam and air in certain hearths to aid in the regeneration of the carbon.

As the carbon is discharged from the furnace, it drops into a quench tank. This is just a small tank filled with water up to a level set by a level controller or float valve. The discharge from the furnace is done through a stainless steel chute, which can either be submerged in the water or fitted with a water curtain. In both cases, isolation of the red hot carbon from the air is obtained. The quench tank can be fabricated from carbon steel, or if iron contamination is objectionable, stainless steel, stainless steel clad, epoxy lining, or resin can be used then. Fines floating in the water are discharged through an overflow to drain.

Off-gas Equipment

To convey and dispose of the hot gas leaving the furnace through the top exhaust connection, the following equipment is needed:

- ductwork and dampers
- afterburner
- wet scrubber
- dust collector
- fan
- stack

Ductwork as well as the damper used to regulate the flow and the pressure in the furnace are usually made of stainless steel.

The afterburner is used in those installations where obnoxious gases are driven off the carbon or formed during oxidation of the impurities. This is just a refractory-lined chamber fitted with a burner and excess air injection

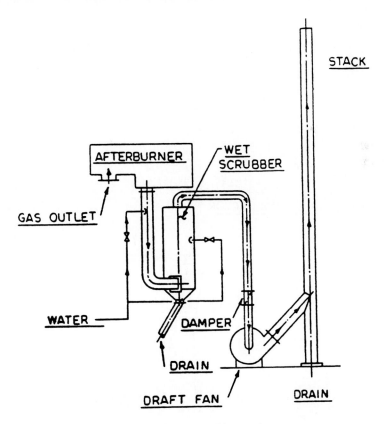

Figure 6.8   Afterburner-scrubber-stack sequence.

suitable for burning at high temperatures. It is installed right after the furnace as shown in Figure 6.8.

A wet-type scrubber is normally used after the afterburner, not only to collect the dust which cannot be burned in the afterburner but also to cool down the gases so that the induced draft fan can handle them. The scrubber is fabricated of stainless steel against corrosion and abrasion by the wet carbon dust. Due to more stringent air pollution regulations, the three-stage impingement-type scrubber has lately been recommended, and in the future it might be necessary to go to the high energy-type scrubbers, such as the Venturi.

In some cases where air pollution regulations are not too stringent but particulate collection is desirable, a dust collector is used. This collector is also fabricated of stainless steel, as shown in Figure 6.9.

To draw the gases from the furnace and through the afterburner, scrubber, and so on, an induced draft fan is necessary in the system. The fan can be fabricated of stainless steel or with a cast-iron casing and stainless rotating parts. Usually, a variable-speed drive is furnished with these units.

The stack to discharge the gases is typically 30–40 ft high and can be made of carbon steel or stainless steel and provided with a clean outdoor drain connection.

The system uses automatic controls on many of its functions and can be automated to the point where the whole operation can be programmed

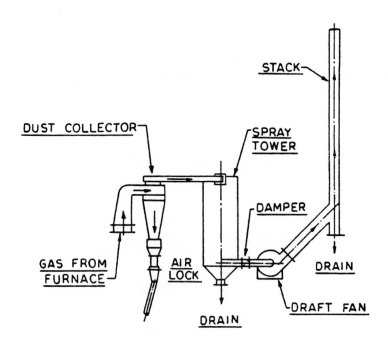

Figure 6.9   Particulate removal-spray tower-stack sequence.

through a computer. However, automatic controls are still deficient, controlling the level in the slurry receiving tanks. Controls used are as follows: (1) mechanical instruments, such as the bindicator type; (2) electric controls, such as the capacitance type; and (3) pressure-sensing elements and others have been used, but with limited success. The operation is not reliable and repetitive and causes numerous malfunctions in the overall operation of the system. The difficulty lies in gauging accurately the location of the solid-liquid interface in the slurry. The use of these controls is naturally in the dewatering tank, surge tank, and quench tank.

These are used in the furnace proper to control the temperature in the fired hearths of the furnace or in the afterburner chamber. They can be of the floating on-off type of proportional positioning. The control signal energizes an electrical or pneumatic actuator that positions the butterfly valve in the air supply line to the burner. The fuel which is proportioned to the air by a special regulator will vary with the air pressure, and the heat input is thus matched to the heat requirement to keep the temperature at the set point.

Another important set of controls normally furnished with the furnace deals with the burners. By their use, the pilots are ignited, the flame is monitored through electronic relays, the system checked and, if found safe, the burner valve is energized and opened. This system utilizes the ultravision flame scanner. The scanner is electrically interlocked with safety valves which are normally assembled in a definite pattern suitable for approval by the Factory Insurance Association (FIA) or Factory Mutual (FM).

Another safety system is the FM system. But this one does not have any flame-scanning device. It only assures safety during the light-off period by making certain that all gas valves have been closed before the main safety shut-off valve is opened. However, once the burners are on, there is no protection from a flame-out. In a carbon furnace where temperatures are usually 1,600–1,800°F, this is acceptable as relighting of the burner will normally occur at those temperatures.

The motor controls are usually mounted in the main control panel for convenience of operation.

In addition to the temperature controllers, a multipoint temperature recorder is also furnished to record the temperatures in all the furnace hearths, shaft cooling-air outlets, the combustion gas temperatures after the afterburner, before the scrubber, and after the scrubber.

The draft, especially that available in the bottom hearth, is also indicated by a gauge so that the outlet damper can be adjusted and the furnace kept at a slightly positive pressure. This will effectively prevent leakage of air into the furnace and burning of the carbon.

There is a normal sequence of safety interlocks and alarms built in the control system which will signal and/or shut the furnace down on such conditions as low or high gas pressure, low combustion air pressure, loss of draft, furnace shaft stoppage, power failure, high temperature, burner flame-out, slurry pump failure, feeder stoppage, and so on.

Operating Conditions

Operating conditions normally encountered in carbon installations affect the furnace operation and degree of regeneration.

Not less than two hours after dewatering, the carbon has been found to have between 45 percent and 50 percent moisture. It has also been found that in a closed vessel, the moisture will stay at this range even after an eight-hour dewatering period.

The normal operating temperatures in the regeneration zone will be 1,650°F to a maximum of 1,850°F. Above the upper fired hearth, the temperatures, of course, decrease rapidly toward the top hearth where the exit temperatures are usually in the order of 600–700°F.

The normal retention time is about 30 minutes, with about 15 minutes required at the regeneration temperature.

The excess air used for the selective oxidation of the organic impurities is on the order of 20 percent of the amount required for burning at stoichiometric conditions. This air can be introduced either through the burners as excess air, or by means of air nozzles welded to the furnace shell, and through the refractory discharging the air in the unfired hearths.

The use of steam to aid in the regeneration of carbon is still of debatable value as the impurities' characteristics determine whether their ready distillation and oxidation can be aided by steam addition.

There are instances in which steam is definitely needed to accomplish the degree of regeneration required. There are other instances where it actually does not make any difference one way or the other. When steam is used, it has been found that a proportion of one pound of steam per pound of regenerated carbon is more than ample.

The afterburner should be operated at a temperature suitable for the ignition and burning of the hydrocarbon components given off by the distillation of the organic impurities. Normally, this would range between 900°F and 1,400°F, and in some cases up to 1,800°F. Plenty of excess air (50 percent) should be available to accomplish this burning.

Inside the furnace, the pressure should be kept slightly positive, somewhere between 0 and plus 0.05 inches W.C. This will prevent leakage of undesirable excess air into the furnace, which would burn the carbon.

Carbon losses in the furnace are difficult to measure. Carbon losses in the whole system are not accurate but are easier to measure by taking the filter-bed height before and after regeneration. The losses, including carbon burned or lost as dust, may range anywhere from 5 percent to 10 percent depending on the installation. However, some older installations where experienced personnel are employed are known to have losses as low as 3 percent.

There are several ways of measuring the degree of regeneration of the carbon. However, none is 100 percent reliable and essentially all it does is give a comparative condition to use as a guide. The only true evaluation is the actual performance of the regenerated carbon under process conditions as compared to the virgin carbon. Parameters can then be set using such physical characteristics as iodine number, molasses number, adsorption isotherms or apparent density, and the relationship to one another to serve as operating controls. Of course, the easiest is the apparent density measurement which only takes a few minutes and, if properly correlated to the other characteristics and to actual performance, can be a very valuable tool.

After the quality control has given an indication of the degree of regeneration attained, it will be necessary to change certain process variables to influence this degree one way or the other. The available variables are listed in the following paragraphs.

The temperature has a definite influence in the degree of regeneration. The higher the temperature, the more the carbon will be regenerated. However, it is also true that the higher temperature will also affect the carbon particles adversely so the usual procedure is to operate at the lowest possible temperature that will give the best product.

Air, of course, is needed to burn off the impurities; and naturally, industrial installations do not behave like laboratory equipment. Therefore, to ensure that all impurities are driven off, enough air has to be introduced. This, of course, burns some of the carbon. However, by manipulating the excess air availability up or down, the degree of regeneration can be controlled.

The comments in a previous paragraph would apply here and, as stated before, steam injection will in some cases affect the degree of regeneration.

The longer the carbon is in the furnace, the higher the regeneration. The retention can be varied by increasing or decreasing the shaft rotational speed. Normally, this speed is on the order of 1 rpm.

ECONOMICS

Factors that influence the economy of a regeneration operation are:

- Type and quality of water treated by the carbon, whether sewage water, industrial wastewater, and river or lake water.
- Losses in the regeneration system.
- Experience of personnel.

The first factor will influence the amount of carbon that will be used per million gallons of water. This may vary from say 300 lb/million gallons up to 1,000 lb/million gallons. This relationship will give the amount of carbon to be regenerated.

The second factor to be estimated depends not only on the furnace itself but also on the physical layout of the plant and how well the carbon is handled. Losses may vary from a minimum of about 3 percent to a maximum of 10 percent.

The third important factor is in those installations where the personnel have become proficient in the operation of the furnace and its related equipment. Such facilities have undoubtedly very low losses and, consequently, lower operational costs.

REGENERATION IN A MULTIHEARTH FURNACE

Figure 6.10 shows a cross section of a multihearth furnace. Since the equipment is well known, only a few brief points will be made about mechanical and process aspects peculiar to this type of equipment.

The multihearth furnace is a device to accomplish heat and mass transfer between gases and solids, passing the gases and solids countercurrently through a series of compartments or stages. In each of these stages, the gas travels in mixed laminar flow over solids spread in thin furrowed layers that are periodically raked to both mix the solids and advance them through the compartment.

Heat transfer is both by direct convection and radiation from the gas to the thin widespread area of solids, and by indirect transfer to the brick walls and parallel brick compartment roof followed by reradiation to the solids, plus minor amounts of conduction through the hearth floor from the gases of the next compartment below. Mass transfer is solely by convection as the gases pass over the solids in laminar flow.

Burners, supplying oxidizing or reducing products of combustion of fuel burned, nozzles injecting steam, or air-injection nozzles apart from fuel burners can be placed at any hearth. Thus, temperature and atmosphere can be changed quite sharply from one hearth compartment to the next. If the carbon reaction zone covers two or more hearths, early reaction while the carbon is heavily laden with residue from adsorption may be carried out at one temperature and atmosphere. The final reaction as the carbon nears the regenerated state may be carried out under a different atmosphere or temperature.

As to solids flow, it is apparent that the carbon is spread relatively thinly over a large area and stirred periodically. For a given furnace, and at given volumetric feed rate, the faster the center shaft is turned, the shorter the retention time. The conveyance of material through the furnace is by positive displacement. All particles have equal retention time except for some slippage and/or some short-circuiting, giving about a ± 15 percent range to particle retention time under conditions of normal bed depth.

The faster the center shaft is turned, the more frequent the stirring and the shorter the interval in which particles lying on top of the bed remain there. Up to a three-minute interval this time between stirrings has not been grossly or obviously significant in reactivation results. Neither has the effect of intervals between stirrings been really fully investigated.

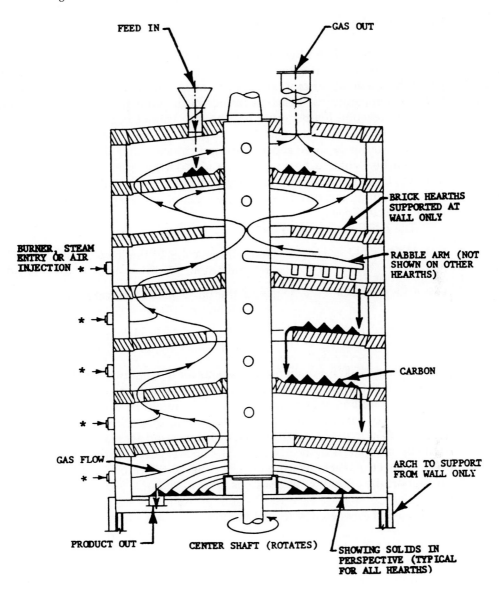

Figure 6.10 Multihearth furnace.

There is, however, a danger in turning the shaft too fast for a given feed rate. The average bed depth becomes too thin to completely cover the distance between one rabble tooth and another. The rabbled bed of carbon is left in furrows or corrugations. The interface between carbon and gases, which is the determinant of capacity, is actually 1.2–1.4 times the nominal or plan-view hearth area, as the angle of repose of the carbon goes from 30° to 45°. The rabbling action itself is extremely gentle.

Reactivation Process

It is usually considered that thermal regeneration of carbon occurs at 1,500–1,850°F by reaction of water vapor and/or carbon dioxide with whatever is left of the adsorbate after it, along with the original carbon, has been heated to these temperatures. Reaction cannot begin until the material has been heated to reaction temperature, and heating cannot begin until the carbon has been dried. Thus, there are three steps which may, with best design and operation, occur in three separate furnace zones rather than overlap. They are a reaction zone, a heating zone, and a drying zone, considering them from the bottom of the furnace upward, following the path of the gas flow, and also naming them in their probable order of importance rather than their order of occurrence.

The reactions expected are:

$$C + H_2O \rightarrow CO + H_2 \qquad \text{(endothermic 4,800* Btu/lb carbon)}$$
$$\text{K equilibrium 1,700°F} = 47.8 \qquad (6\text{-}1)$$

$$C + CO_2 \rightarrow 2\,CO \qquad \text{(endothermic 5,950* Btu/lb carbon)}$$
$$\text{K equilibrium 1,700°F} = 52.5 \qquad (6\text{-}2)$$

$$CO_2 + H_2 \rightarrow CO + H_2O \qquad \text{(endothermic 37.3 Btu/ft}^3 \text{ any constituent)}$$
$$\text{K equilibrium 1,700°F} = 0.73 \qquad (6\text{-}3)$$

Even at 1,500°F, equilibrium constants for the first two reactions are high enough (about 10) to expect reaction to go essentially to completion except for kinetic-rate limitations. The reaction zone might be expected to be sized by volume of rabbled carbon bed, considering that the carbon gasification reactions that occur in it are governed by kinetics and are reaction-rate limited. Actually, it is sized by hearth area. The area exposed to the gases controls mass transfer of reactants from the gas phase to the carbon and heat transfer to support the endothermic reactions.

Fluidized-Bed Powdered Activated Carbon

The use of activated carbon for the tertiary treatment of secondary sewage effluents has been used extensively. A schematic diagram of the process is shown in Figure 6.11. Powdered carbon is as effective as granular activated carbon for removing the organic impurities from the wastewater.

Before powdered carbon can be used commercially or reused for tertiary treatment of sewage effluents, a method of regeneration is required. The use of the fluidized bed for regeneration offers the key advantages of excellent temperature and atmosphere control and the ability to process the powdered solids conveniently and continuously. However, the median diameter of the

---

* Assumes carbon is graphite, whereas most of the reactions in regeneration are with a modified form of amorphous carbon.

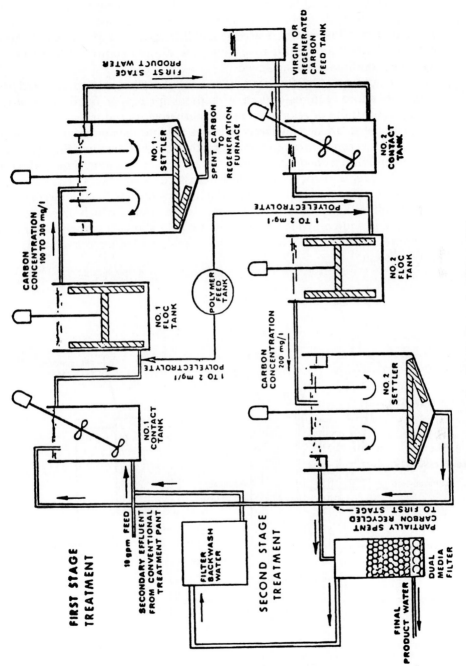

Figure 6.11 Powdered carbon adsorption system.

161

carbon particles is approximately 11 μ, which is considerably finer than normally used in fluidized-bed operations. The problems associated with fluidization of very fine powders are the inability to achieve proper fluidization and the high entrainment losses. Thus, the desired control of temperature and retention time is not achieved.

The application of fluidization methods for regeneration of powdered carbon therefore requires operating methods for proper fluidization of the fine powder or the development of an alternative procedure.

The spent, dried, powdered carbon is fed into the bottom of the coarse, inert bed and carried through the bed by the action of the fluidizing gas. This method of fluidized-bed operation offers control of retention time of the fine carbon powder and shows good heat-transfer characteristics. The finely divided carbon is recovered from the effluent gas stream with cyclone collectors or some other collection device. A bench-scale system which illustrates this technique is shown in Figure 6.12. Figure 6.13 shows a carbon adsorption solvent recovery system with an in situ steam regeneration operation as part of the system.

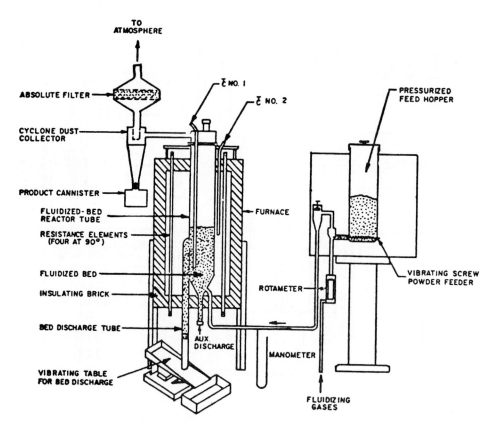

Figure 6.12 Bench-scale fluidized bed and auxiliary equipment.

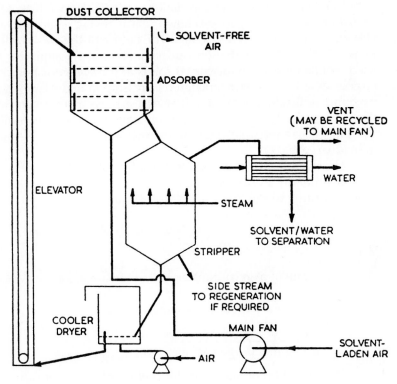

Figure 6.13 A carbon adsorption solvent recovery system with an in-place steam regeneration operation.

Wet-Oxidation Regeneration

Wet oxidation may be defined as a process in which a substance in aqueous solution or suspension is oxidized by oxygen transferred from a gas phase in intimate contact with the liquid phase. The substance may be organic or inorganic. In this broad definition, both the well-known oxidation of ferrous salts to ferric salts by exposure of a solution to air at room temperature and the adsorption of oxygen by alkaline pyrogallol in the classical Orsat gas analysis would be considered wet oxidations. Most applications of commercial significance require some elevation of temperatures and pressures. A range of about 125°C (257°F) and 5 atm to 320°C (608°F) and 200 atm covers most cases. Frequently, air is the oxygen-containing gas, in which case the process may be termed *wet-air oxidation* (WAO). In the general case, including the use of pure oxygen, the broader term of *wet oxidation* (WO) is used. WO may be controlled for a specific synthesis step, held at intermediate degrees in some applications, or forced to completion for waste disposal.

Variables that affect wet-oxidation rates are temperature, total pressure, partial pressure of oxygen, gas-liquid phase mixing, concentration of oxidizable substances, pH, and catalysts. The nature of the waste will often determine whether a concentration can be altered, a pH adjusted, or a catalyst used. Solubilities of any dissolved solids at the system conditions must also be considered. Another restriction on pressure in many cases is the need for maximum utilization of oxygen. If an excess pressure of oxygen is maintained throughout the liquid-phase detention time, some oxygen will be wasted. The most cost-effective oxygen partial pressure must be considered for each individual case.

The overall rate of oxygen transfer from the gas to the liquid phase is limited primarily by the resistance to mass transfer in the liquid-diffusion layer according to the following classical equation:

$$r = K k_1 a (C - C^*)$$

where

r = rate of oxygen transfer per unit volume, kg mol/M^3/sec or lb mol/ft^3/hr

K = oxygen solubility constant in water,

$$\frac{\text{kg mol oxygen/}M^3 \text{ water}}{\text{kg mol oxygen/}M^3 \text{ air}}$$

k_1 = liquid film mass-transfer coefficient, m/sec or ft/hr

a = interfacial area per unit volume, M^{-1} or ft^{-1}

C = oxygen concentration in the gas phase, kg mol/m^3 or lb mol/ft^3

C^* = the hypothetical oxygen concentration in the gas phase that would be in equilibrium with the liquid phase, kg mol/m^3 or lb mol/ft^3

The solubility of oxygen in water, K, increases strongly with temperature and may to some extent be affected by the presence of substances dissolved in water. The liquid film mass-transfer coefficient, k_1, depends on the diffusion coefficient of oxygen in water, the concentration and type of substances dissolved in water, the intensity of stirring in the liquid phase, and the presence of the chemical reaction increasing greatly with increases in temperature. The interfacial area, a, depends on the liquid surface tension, the liquid physical properties, the gas-phase flow rate, the intensity of stirring in the liquid phase, and the concentration and type of substances dissolved in the aqueous phase. The interfacial area also increases greatly with temperature.

Because of the difficulty in separately measuring the liquid film mass-transfer coefficient and the interfacial area, these two quantities are most often correlated together as the volumetric mass-transfer coefficient, k_1a.

As shown by the equation, the solubility of oxygen in water can have a strong influence on the rate of oxygen transfer from the gas phase to the liquid phase. The solubility increases strongly with temperatures over the range of temperature; above the critical temperature and pressure of a particular air-water mixture, the solubility is considered infinite. Though the solubility of oxygen in water at ambient conditions may be somewhat affected by the presence of dissolved substances, its effect in aqueous solutions at wet-oxidation temperatures and pressures is expected to be very minor when all factors are considered.

REGENERATION BY STEAM

Adsorbed solvents and some organics may be stripped from activated carbon by means of direct steam. This method is essentially an in-place regeneration of carbon in the on-line adsorber. The comparison of several low-temperature methods of regeneration can be compared with activated carbon charged with 20 percent ether:

Regeneration method	Percentage of charge expelled
Heating at 100°C (212°F)	5
Vacuum 50 mm Hg at 20°C (68°F)	10
Gas circulation at 130°C (266°F)	20
Direct steam at 100°C (212°F)	100

The stripping process is a dynamic process and the quantity of steam required to desorb a given amount of carbon is dependent on the physical design of the adsorption plant rather than on the theoretical amount of heat required to distill the solvent from the carbon. Care must be taken with the design of the steam manifold, carbon-bed distillate piping, and condensing system to achieve an efficient steam consumption. A well-designed plant will have a steam consumption in the region of 1–4 pounds steam per pound of recovered solvent. The following factors must be considered when designing the stripping process.

- The length of time required for steaming or regeneration should be as short as possible, as this represents time off from adsorption. If continuous recovery is to be maintained, a standby vessel must be used to recover solvent while one vessel is undergoing regeneration. Generally two or more adsorbers are used to permit continuous recovery.
- The shorter the time of regeneration, the higher will be the steaming rate, and greater will be the heat duty of the condensing system. In order to keep the size of the condenser efficient, the regeneration times are usually chosen between 30 minutes and 60 minutes.
- The steaming direction should be in the opposite direction to the adsorption. This prevents the accumulation of polymerized substances, if

present, from being driven deeper into the carbon bed and permits the stripped solvents from the upper layers to wash out the lower layers of the carbon, thus assisting the removal of adsorbed substances.

- For rapid stripping and high heat transfer it is essential that the inert gas or air is swept out of the adsorber and condensing system as quickly as possible. The design of the adsorber vessel, steam-manifold, and condenser can greatly assist in the removal of unwanted air.
- Steam economy may be obtained by placing the distillate piping and condenser below the adsorber to enable the condensed liquid in the adsorber to drain directly into the condenser and separator. This prevents accumulation of liquid in the distillate piping, which would obstruct the free path of air and steam to be swept through the system.
- The quality of the steam is an important factor. A high proportion of the heat content of the stripping stream is used to heat up the adsorber vessel and its carbon contents. To achieve rapid heat transfer, it is important that the steam condense quickly. The steam should, therefore, contain only a slight superheat to allow condensation to take place.
- The quantity of steam will depend on the physical properties of the adsorber and its carbon contents.
- The carbon should have a low retentivity to enable the solvent to be easily stripped out.
- Stripped solvents are recovered by condensing and cooling the steam-solvent mixture removed from the carbon bed. Tubular heat exchangers can be used for the condensers and aftercoolers.

During the steaming period, a certain amount of condensation takes place in the carbon bed, increasing the moisture content of the carbon. The amount of moisture may be controlled to a certain extent by the dryness of the steam used and the pressure and temperature conditions in the carbon bed during steaming.

In order to achieve the maximum adsorption capacity and adsorption efficiency, when recovering miscible solvents, it is desirable that the carbon bed should be dried and cooled before being returned to the solvent air stream. The desired state of dryness depends on the physical properties of the solvent and concentration of the solvent in the air or gas stream. It is desirable when using high-solvent concentrations to leave some moisture in the carbon so that the heat of adsorption may be used in evaporating the moisture from the carbon, thus preventing an undue rise in the temperature of the carbon bed.

The heat required for drying the carbon after steaming may be supplied by several methods, either singly or in a combination:

- Sensible heat of carbon and adsorber in cooling from steam temperature down to adsorption temperature.

- Sensible heat content of a heat generator placed below the carbon bed which becomes heated during the steaming period.
- Use of an external air heater.
- Use of the heat of adsorption by recovering from a sufficiently high solvent concentration when drying with solvent-laden air or gas.

Air should not be used for drying if using solvents which may oxidize, polymerize, or decompose when in contact with hot moist activated carbon. The catalytic effect of activated carbon in this condition may cause such solvents to break down or hydrolyse, causing the formation of acids. In these cases it is desirable to dry and cool the carbon with inert gas.

7

Exposure, Controls, and Safety

This section, pertaining to worker exposure to the carbon adsorption unit operation and safety considerations in such operations, begins with a general discussion of the factors which affect the potential exposures to the unit operation. As with other unit operations, there are three modes of operation: start-up, operation, and turnaround. Start-up and turnaround typically involve the most acute exposures (short term, possibly high concentrations of dusts, solvents, and temperature). Turnaround is generally meant to include cleanout, carbon additions, bed replacements, process changeover, and so on. Quantification of the total exposure to a carbon adsorption operation must involve properly weighing the individual exposures from each of these modes with respect to carbon dust, volatile organic gases, liquid organics, and temperature.

Aside from the mode of operation, there are many key physical and chemical parameters which influence the potential for worker exposure during gaseous-phase or liquid-phase carbon adsorption unit operations.

- The nature of the process stream. Whether a gaseous-phase of liquid-phase carbon adsorption operation, the volatility and toxicity of the materials in the process stream remain of paramount importance with respect to worker exposure and the need for a thorough program of safety work practices and engineering controls.

- Nature of the components that the operation is designed to remove. These materials will be adsorbed onto the activated carbon and then have a relatively high potential for worker exposure during carbon-bed inspection, carbon additions, carbon replacements, and so on.
- Physical condition of the adsorption equipment. This includes the condition of the carbon and all equipment accessories.
- Location of equipment. Most carbon adsorption equipment is located outside, thus minimizing the potential for worker exposures during operation.
- The operating parameters of the equipment. Temperatures, pressures, fluid velocities, and so on of the operation all influence the potential for worker exposure to toxic or volatile materials.
- The method of handling new and spent activated carbon. One of the major concerns in carbon adsorption is the exposure to the activated carbon itself. Good work practices during charging, supplementing, and removing spent carbon minimize this potential exposure area.
- Air should not be used for drying if using solvents which may oxidize, polymerize, or decompose when in contact with hot moist activated carbon. The catalytic effect of activated carbon in this condition may cause such solvents to break down or hydrolyze, causing the formation of acids. In these cases it is desirable to dry and cool the carbon with inert gas.

Details should be provided with as much information as possible concerning the process and manner of working in order to provide the plant best suited to the process conditions. A suggested list of items to be included are:

- Name and chemical formula of the solvents to be recovered. If a mixture of solvents is to be recovered, it is necessary to know the proportions either by weight or volume.
- Condition in which the solvent is to be recovered (if a mixture of solvents is to be recovered, whether these have to be separated into individual components). Also the amount of water which may be tolerated in the recovered solvent.
- Characteristics of the solvent (boiling point, specific gravity, solubility in water, azeotropes, special hazards, and so on).
- Quantity of solvent normally used per hour, the number of hours worked per day, and number of days per week, or plant data on raw material, feeds, and finished products from which the solvent rate may be calculated.
- Variations in the rate of solvent evaporation.
- Whether solvent is evaporated in an air or gas stream and if gas, the composition of gas used.

- Volume rate of air or gas used to evaporate the solvent.
- Concentration of solvent in the air or gas stream, and the desired limit of solvent concentration in the air.
- Temperature and pressure of the solvent-air mixture, either at machine or where it is proposed to enter recovery plant.
- Humidity of air or gas stream.
- Description of material in the solvent and whether there is any contaminant being entrained in the air stream and carried to recovery plant (paper, dust, fluff, dirt, foreign matter) which could clog the fine pores of the activated carbon.
- Description of the process, type of equipment, and contributions to the total of any individual sources of solvent vapor.
- Whether the process is enclosed and complete with solvent air ducts, flame traps, controls, and so on for collecting solvent vapors from every evaporation source or outdoors.
- Services including cost and quality of steam and water supplies.
- Site area, head room, enclosed outside location, and distance from solvent evaporation process or source.
- Possibility of hanging the solvents to be recovered at some future date or future expansion.
- Method of operation or plant—whether manually-operated valves or automatic controls.

The routes of exposure for either liquid-phase adsorption or gaseous-phase adsorption are dermal, through splashing of liquids or contact with the activated carbon, and inhalation, through equipment leaks and exposure to pulverized carbon dust.

Carbon adsorption process of solvent recovery may be applied to practically all of the solvent-using industries. The nature of the process, in general, involves highly flammable or toxic substances. It is essential, from a safety point of view, to insure that the ventilation system and the recovery plant are designed and installed properly.

The risk of fire, with its accompanying disastrous effects, cannot be overemphasized when using flammable solvents.

All of the usual safety precautions necessary when handling toxic or flammable solvents must be strictly observed. Recommended methods of handling solvents are given in safety codes and regulations dealing with the storage of toxic chemicals and solvents. The collection of solvents from evaporating sources requires:

- Adequate ventilation to prevent solvent concentrations from reaching dangerous proportions in work areas or equipment.

- Adequate hooding and ducting to prevent solvent vapor polluting the work area where people are employed.
- Isolation of equipment and sources of solvent by the fitting of valves and flame traps in the ducting between solvent vapor-collecting devices. This helps to localize and extinguish any accidents which may occur.

All types of activated carbon if raised to a suitable temperature and exposed to oxidizing conditions will burn. This tendency, however, depends on the quality of the carbon employed, and this may vary considerably. The majority of activated carbons are safe for operation at air temperatures below 150°C (302°F). Temperatures during normal operation of a solvent recovery plant include:

1. During the adsorption period, the temperature of the carbon may be anything from room temperature up to 50°C (122°F).
2. During desorption, that is, the period when saturated steam is passed through the adsorber, the highest temperatures are usually from 120°C to 140°C (248°–284°F).
3. During the drying of the carbon, the temperature is reduced owing to evaporation of moisture from the carbon, falling from 110°C to 40°C (230°–104°F).

Safety Considerations

The carbon adsorption process of solvent recovery may be applied to practically all of the solvent-using industries. In view of the nature of the process in that it generally involves dealing with highly flammable or toxic substances, it is essential from a safety point of view to insure that the ventilation system and recovery plant are properly designed and installed. Too little knowledge may lead unwittingly to the design of a hazardous plant, and great care must be taken to ensure that the final installation is as safe as possible.

The risk of fire, with its accompanying disastrous effects, cannot be overemphasized when using flammable solvents. Safety precautions for such materials should be strictly adhered to. Activated carbon has a strongly catalytic effect on certain solvents, which are oxidizable at the temperatures at which a carbon adsorption plant may operate. When commercial solvents such as benzene, toluene, xylene, alcohol, ether, acetone, and common esters are recovered, no difficulties should be experienced providing the recovery plant is properly designed and operated. There is no risk of oxidation during the adsorption period providing:

- Temperature of the carbon is limited to between 20°C and 50°C (68°–122°F).

- A low-solvent concentration is used so that the thermal capacity of the air is very great in relation to the heat of adsorption.
- The heat of adsorption is so small that even in the most favorable case the highest temperature rise during an adsorption period is limited to 10°C (50°F).
- The carbon is dried and cooled after each steaming before being connected onto the solvent air stream. Other safety features should include:
  - Use of a high-quality activated carbon of regular grain type.
  - Use of valves that insure long life and freedom from leakages.
  - Progressive drying of the carbon after steaming at the lowest possible temperatures.
  - Sufficient instruments, such as thermometers in the carbon bed and flow meters for measurement of solvent air rate, to ensure that the plant operator is in full control of the adsorption plant and is quickly informed of any change from normal operation.
  - Proper instruction for plant operators. Trained operators and detailed working instructions not only cover the essential precautions to secure recovery efficiency but also to maintain safety.

Labor Requirements

The analysis of potential worker exposures from liquid-phase adsorption must cover a wide range of scenarios from batch operations to continuous liquid-phase adsorption operations. It is expected that the labor requirements for batch-operated, liquid-phase adsorption equipment are far greater than those required for the supervisory control of continuous liquid-phase or gaseous-phase adsorption. Batch liquid phase probably requires one operator shift to charge the process vessel with feed solution, add the activated carbon and operate the dust control system, operate the agitator for a predetermined time cycle (which could range from 30 minutes to 2 hours), pump the mixture through a filter to remove the carbon, wash and dry the filter cake and discharge, and dispose of the activated carbon.

The magnitude of the operations, that is, the number of batch adsorption units, complexity of equipment, and the complexity of the operation will of course influence this labor estimate. In actual applications the total labor demand would range from the one employee per shift estimate already discussed to numerous direct operators plus various types of supporting personnel. It is also necessary to decide whether an adsorption operation can be isolated from associated (that is, filtration) operations before discussing labor requirements, exposure potential, controls, and so on.

Continuous liquid-phase adsorption requires less direct operator presence than batch liquid-phase carbon adsorption. Many of the operating sequences are automatically controlled. The actual labor requirements are more a function of the total adsorption system design, which includes the associated processing facilities such as filtration. The frequency of the periodic emptying of the adsorption chambers and recharging fresh carbon would be one criterion for determining the need for additional labor. The specific exposure hazards and the need for protective equipment and engineering controls are again determined by the material being processed along with the operating conditions.

An example of a continuous liquid-phase adsorption system involved in a total system would be a sugar-processing application. In this situation there may be several in-line regeneration furnaces which are used for the periodic regeneration of activated carbon. It is typical for these furnaces to operate at temperatures of approximately 900°C. In any application where regeneration furnaces are part of the total system, additional exposure potentials are presented.

Exposures/Routes of Exposure

The potential exposures from liquid-phase carbon adsorption can be the result of equipment leaks, splash-loading of carbon, contact with exposed filter cake, and so on. The routes of exposure, therefore, are primarily dermal and some inhalation. Both the granular and powdered carbons which are used in liquid-phase adsorption can be dusty when handled improperly. The powdered types present more of a problem. However, the dusting can be minimized. The severity of exposures depends on the physical and chemical properties of the solution being treated, such as pH, toxicity, and so on, as well as the processing conditions, such as temperature. The use of sound operating practices and protective equipment consistent with the types of materials being processed minimizes the possibility of worker-exposure problems. Additional attention to the method used in loading fresh carbon and removing spent carbon minimizes the possibilities for any dust exposure from carbon handling or liquid exposure from splashing.

Protective Equipment/Engineering Controls

In small plant operations where batch fixed-bed liquid-phase adsorbers are used in series or in parallel, the loading and unloading of carbon are potentially very messy operations. Personal protective equipment in this situation includes full-face respirators equipped with HEPA cartridges for protection against inhalation exposure and a full suit of disposable protective clothing, including gloves for dermal protection for carbon-loading operations in which the carbon and the liquid are charged manually or not under enclosed/evacuated conditions. Regardless, fugitive carbon fines are always a problem and cause billowing and puffing during the charging operation. Water sprays are commonly used to control fugitive carbon dust. They can either be incorporated into the system or used in the form of an add-on water hose.

One method of controlled handling available to larger users of powdered carbon is bulk handling. Powdered carbon arrives via bulk rail car or truck and is pneumatically conveyed to storage, dry or as a slurry. The closed slurry tank should be under a slight vacuum and the carbon should be added to the liquid as fast as it can be wetted. The dust can be exhausted to an appropriate collection system. The use of a powdered carbon/water slurry feed or premoisturized carbon has greatly improved the ease of handling powdered activated carbon.

It is also common to have the proper combination of chemical and dust full-face respirator cartridges plus protective clothing and gloves for the removal of spent carbon. The concern, of course, is to protect the worker from contact with the material that was adsorbed, such as a toxic sludge for liquid-phase operations. The selection of the specific protective equipment to be used should focus on the hazards of the particular materials present in the carbon.

GASEOUS-PHASE ADSORPTION

In most applications, continuous gaseous-phase carbon adsorption requires only the supervisory attention of an operator. Most gaseous-phase operations are operated continuously with automatic cycling of the fixed activated carbon beds between active adsorption, regeneration, and cooling (for three-bed systems). When the direct labor requirement is more than a strictly supervisory role, the need for more active involvement by the workers has been determined by the overall system design, size, number of adsorption units, the cycling control procedure (operates on timer or on instrumentation), and sampling or inspection frequency. There may also be a need for additional supporting personnel based on the need of the total system.

The potential routes of exposure from gaseous-phase carbon adsorption systems are dermal and by inhalation. The potential sources of exposure are equipment leaks, high temperatures, and handling of the activated carbon. The handling of activated carbon during charging, additions, and removal potentially exposes the worker to fine carbon dust and toxic dust that is adsorbed on the carbon.

Work Practices/Protective Equipment/Engineering Controls

The same protective equipment discussed for liquid-phase adsorption (that is, respirators with appropriate particulate/organic cartridges and disposable protective clothing and gloves) may be advisable for certain gaseous-phase adsorption applications. Again, the severity of the exposure potential must be considered on a case-by-case basis with respect to the types of materials being processed and the operating conditions present such as temperature and pressure. For many applications, the only time when a significant exposure potential would be present is during worker contact with the carbon beds during replacement or additions. Where considerable amounts of solvents are handled, additional provisions (that is, ventilation) are employed.

It is possible that certain gaseous-phase adsorption systems incorporate some type of fines-removal system (fabric filters, cyclones, venturi scrubbers) to control fugitive particulates from powdered activated carbon handling.

Each of the possible regeneration processes poses individual exposure potentials and needs to be evaluated separately. Regeneration operations are usually dictated by process economics. The options range from discarding spent carbon after once-through use, to a secured landfill, to regeneration of the activated carbon on-site by desorbing the adsorbed materials through vaporization and condensation, to thermal processes for volatilizing and combusting the adsorbed materials.

The following are among the good safety work practices which should be adhered to in operating a gaseous-phase carbon adsorption system. Employees should never enter any system where toxic or explosive gaseous mixtures are present. As part of the unit operation, it is important to guard against the existence of any explosive combinations during purging or cooling cycles. Respiratory as well as dermal protection (organic respirators and full suit of protective clothing) are the proper safeguards for workers who are in close contact with toxic-recovered solvents.

Because some gaseous adsorbents such as activated carbon are flammable, great care should be exercised in desorbing all volatile vapors prior to the introduction of air which could result in premature expenditure of the adsorbent and possible ignition. Additional fire hazards are posed by operations in which lint is captured in filters or where deposits of accumulated organic compounds occur. These are alleviated by frequent cleaning.

Electrical hazards can be minimized by adequate grounding to prevent static electricity buildup which may lead to an explosive situation. Gaseous-phase carbon adsorption systems which are operated under pressure should have proper pressure relief valves and proper vents.

Safety precautions of carbon adsorption equipment for vapor control at petroleum marketing locations include the following considerations. There is always some degree of explosion hazard when dealing with hydrocarbon vapors and air. A carbon adsorption vapor-recovery unit typically operates with a combustible vapor mixture entering the unit and a noncombustible exhaust from the unit. Extreme caution should be taken to vent and/or remove combustibles from the system prior to entering the equipment for repairs, carbon-bed additions, or replacements. Industry personnel are urged to contact the manufacturer prior to entering the carbon beds.

Although there is a minimal amount of moving equipment in this type of carbon adsorption system, operators and plant personnel involved with the machinery should be adequately trained and alert at all times to avoid potential injuries. Employees should be aware of proper shutdown and lockout procedures prior to performing any maintenance on the system. Only fully qualified electricians should work on the electrical equipment associated with carbon adsorption and only after proper deenergizing procedures and lockout. All skid-mounted equipment that is factory grounded to the skid should also be adequately grounded to a ground rod or field prior to switching on the power.

Labor and Supervision

Operation of the adsorption system requires minimal operator attention to record flows, collect samples, and transfer carbon. At most plants, 0.5–1 man-year is sufficient for operating the adsorption system. Operation of the storage and reactivation systems requires greater operator attention. On furnaces smaller than 9,000 kg/day (2,000 lb/day), operator attention is required 4–8 hours per shift. On larger furnaces, an additional operator per shift is usually required to assist in carbon transfers and to provide general plant surveillance. An allowance of 25 percent of the direct labor cost is recommended for supervision.

Maintenance is a function of the value of the installed equipment and the corrosivity of the spent carbon and the wastewater. Maintenance cost of the adsorbers and related piping is typical of chemical processes, and ranges from 5 percent to 7 percent of the installed cost per year.

Maintenance of the storage and handling systems and the reactivation system is more complex, as would be expected for slurry-handling systems and high-temperature equipment. Maintenance costs for this equipment may range from 7 percent to 10 percent of the installed cost per year. Major maintenance items include hearth replacement, furnace arm and teeth replacement, kiln replacement, kiln flighting, brick replacement, pump repairs, and so on. The cost of these items can be reduced through proper selection of materials, continuous furnace operation, and good equipment design. The furnace is a heavy-duty piece of equipment operating at high temperatures. It can be quite reliable provided it is not subjected to frequent thermal shock caused by feed interruptions and shutdowns.

Utilities

Fuel for the furnace and afterburner is the primary utility cost in granular activated carbon reactivation systems. Fuel consumption for different types of furnaces and spent carbons will vary. A good rule of thumb is 2,200 kcal/kg (4,000 Btu/lb) of carbon for reactivation and another 2,200 kcal/kg (4,000 Btu/lb) for incineration of the off-gases. Both natural gas and fuel oil can be used to fire multihearth furnaces and rotary kilns. Heat recovery is generally not practiced due to the corrosivity of the off-gases and the presence of particulates. In most cases, steam is added to the furnace at a rate of one part of steam per one part of carbon to promote selective oxidation of adsorbed organics. Costs for the adsorption system are minimal. The reactivation system and handling systems employ slurry pumps, compressors, combustion air blowers, I.D. fans, and cooling-air blowers, which add to the power costs.

Monitoring and General Plant Overhead

Monitoring costs for on-site reactivation systems include monitoring the adsorption system performance and quality control on the reactivation system. These costs vary from plant to plant depending on discharge permit requirements and size of the operation. A monitoring cost allowance of one man-year is typical of larger systems employing on-site reactivation.

Service fees are a function of the type and concentration of contaminant, effluent objectives, stream flow, plant location, and plant operating schedule. Generally, the adsorption service fee is lower than the annual cost of a custom system employing either on-site reactivation or off-site custom reactivation. This is possible because of the economies that result from proved modular engineering, quantity equipment purchasing, and the economies of scale for reactivation equipment and operation.

Instrumentation for Process Control and Safety

An activated carbon adsorption plant may be controlled quite simply with the aid of flow meters and thermometers. The most important factors for control are:

- *Solvent concentration.* This must be kept at a safe concentration which may be calculated from a knowledge of the weight of solvent being evaporated per hour and the quantity of air or gas being drawn through the adsorption plant. In the majority of cases the rate of solvent evaporation is known, and an orifice-type flow meter will enable the concentration to be calculated and maintained. Low concentrations involved of less than 1 percent by volume may be measured in a number of ways. depending on the physical and chemical properties of the solvent. Methods used include:
  - *Adsorption of the solvent in a small test adsorber containing activated carbon.* A measured volume of solvent-laden air is passed through the adsorber, the carbon is steamed and the solvent condensed, collected, and weighed.
  - *Qualitative analysis.* Chlorinated solvent vapors may be decomposed by passing through a quartz tube heated above 400°C. The chlorine is detected by a 2 percent potassium iodide solution and starch.
  - *Quantitative analysis.* Solvent vapor is adsorbed in a solution and titrated with a selected indicator. Most solvents can be detected and the concentration determined by chemical analysis, but the procedure is often difficult and lengthy, and not very suitable for routine checks on the plant.

- *Infra red gas analyzer, gas chromatography, atomic-absorption spectroscopy.* These techniques have greatly assisted the analyst in detecting and measuring single components of a gas mixture, even from the complex mixture of gases. These methods are capable of detecting a few parts per million of a component in a gas mixture with comparative ease within a few seconds. Analyzers may be used continuously to monitor selected components from one or several process streams such as measurement of solvent concentration at various points, in the plant and building, as well as the solvent concentration at the inlet and outlet of the adsorbers.

- *Temperature of adsorption.* Thermometers should be placed at strategic points in the solvent air line and carbon bed. Any change in method of operation of the adsorption plant is generally reflected as a change in the operating temperature of the plant. Temperatures should be recorded in order that any change will be noticed at once by the operator. The use of recording thermometers in the carbon bed at different levels will indicate the progress of adsorption. This will provide a true record of the operation of the plant and assist in tracing the cause of any faults which may occur, so that they may be rectified with minimum delay and loss of solvent.

APPENDIX

Physical Constants and Conversion Factors

PHYSICAL CONSTANTS OF

In the following tables the more common physical constants are given for hydrocarbons, certain other organic series, and miscellaneous materials. While these constants, in general, are based upon reliable data, estimated values are included in some instances where data are not available.

The values were compiled from a number of sources. The sources most often selected are listed below in approximately the order of frequency of use as a reference.

1. "Selected Values of Physical and Thermodynamic Properties of Hydrocarbons and Related Compounds", American Petroleum Institute Research Project 44, Thermodynamics Research Center, Texas A&M University, College Station, Texas (loose-leaf data sheets, extant 1973).

2. "Selected Values of Properties of Chemical Compounds", Thermodynamics Research Center Data Project, Thermodynamics Research Center, Texas A&M University, College Station, Texas (loose-leaf data sheets, extant 1973).

3. "Technical Data Book—Petroleum Refining", American Petroleum Institute, 2nd Edition (1970).

4. A. P. Kudchadker, G. H. Alani, and B. J. Zwolinsky, Chem. Revs., *68*, No. 6:659 (1968).

5. K. A. Kobe and R. E. Lynn, Jr., Chem. Revs., *52*, No. 1:117 (1953).

6. J. Timmermans, "Physico-Chemical Constants of Pure Organic Compounds", Elsevier Publishing Co., Inc., New York (1950).

7. A. L. Lydersen, University of Wisconsin Engineering Experimental Station, Report No. 3 (April, 1955).

PHYSICAL CONSTANTS OF HYDROCARBONS

| | FORMULA | MOLEC. WT. | BOILING POINT °F | MELTING POINT °F | DENSITY | | | CRITICAL CONSTANTS | | | ACENTRIC FACTOR | HEAT OF VAPORIZ. @ B.P. Btu/lb | HEAT OF COMBUSTION @ 60°F Btu/lb | |
					°API	Sp Gr 60/60	Lb/Gal[a]	t °F	P Atm	D g/ml			GROSS	NET
PARAFFINS														
Methane	CH_4	16.0	-258.7	-296.5	340[b]	0.30[b]	2.50[b]	-116.7	45.4	0.162	0.001	219.2	23882[c]	21503[c]
Ethane	C_2H_6	30.1	-127.5	-297.9	247	0.374	3.11	90.1	48.2	0.203	0.101	210.4	22320[c]	20417[c]
Propane	C_3H_8	44.1	- 43.7	-305.8	147	0.508	4.23	206.0	41.9	0.217	0.157	183.1	21662[c]	19928[c]
n-Butane	C_4H_{10}	58.1	31.1	-217.0	111	0.584	4.86	305.6	37.5	0.228	0.206	165.7	21294[c]	19654[c]
2-Methylpropane (Isobutane)	C_4H_{10}	58.1	10.9	-255.3	120	0.563	4.69	275.0	36.0	0.221	0.188	157.5	21234[c]	19591[c]
n-Pentane	C_5H_{12}	72.1	96.9	-201.5	92.7	0.631	5.25	385.7	33.3	0.237	0.256	153.6	20926	19339
2-Methylbutane (Isopentane)	C_5H_{12}	72.1	82.1	-255.6	94.9	0.625	5.20	369.0	33.4	0.236	0.223	147.1	20887	19300
2,2-Dimethylpropane (Neopentane)	C_5H_{12}	72.1	49.1	2.2	106	0.597	4.97	321.1	31.6	0.238	0.202	135.6	22954[c]	19368[c]
n-Hexane	C_6H_{14}	86.2	155.7	-139.6	81.6	0.664	5.53	453.6	29.7	0.233	0.302	144.0	20784	19232
2-Methylpentane	C_6H_{14}	86.2	140.5	-244.6	83.5	0.658	5.48	435.7	29.7	0.235	0.280	138.7	20756	19205
3-Methylpentane	C_6H_{14}	86.2	145.9	-180.4	80.0	0.669	5.57	448.2	30.8	0.235	0.278	140.1	20768	19217
2,2-Dimethylbutane	C_6H_{14}	86.2	121.5	-147.8	84.9	0.654	5.45	440.0	30.4	0.240	0.236	131.2	20711	19160
2,3-Dimethylbutane	C_6H_{14}	86.2	136.4	-199.4	81.0	0.666	5.55	440.2	30.9	0.241	0.250	136.1	20743	19191
n-Heptane	C_7H_{16}	100.2	209.2	-131.1	74.2	0.688	5.73	512.6	27.0	0.232	0.352	136.0	20681	19156
2-Methylhexane	C_7H_{16}	100.2	194.1	-180.9	75.7	0.683	5.69	494.9	27.0	0.238	0.330	131.6	20658	19133
3-Methylhexane	C_7H_{16}	100.2	197.3	-182.9	73.0	0.692	5.76	504.3	28.1	0.240	0.329	132.1	20668	19144
3-Ethylpentane	C_7H_{16}	100.2	200.3	-181.5	69.8	0.703	5.85	513.4	28.5	0.241	0.314	132.8	20679	19154
2,2-Dimethylpentane	C_7H_{16}	100.2	174.6	-190.9	77.2	0.678	5.65	477.1	27.4	0.241	0.302	125.1	20620	19095
2,3-Dimethylpentane	C_7H_{16}	100.2	193.6	--	70.9	0.699	5.82	507.4	27.7	0.255	0.304	130.4	20642	19117
2,4-Dimethylpentane	C_7H_{16}	100.2	176.9	-182.6	77.5	0.677	5.64	475.7	27.0	0.240	0.307	126.6	20639	19111
3,3-Dimethylpentane	C_7H_{16}	100.2	186.9	-210.0	71.2	0.698	5.81	505.7	29.1	0.242	0.285	127.2	20638	19113
2,2,3-Trimethylbutane (Triptane)	C_7H_{16}	100.2	177.6	- 12.8	72.1	0.695	5.79	496.3	29.1	0.252	0.259	124.2	20627	19103

a. In air, @ 60°F.   b. Hypothetical (extrapolated) value.
c. Heat of combustion as a gas, otherwise as a liquid.

PHYSICAL CONSTANTS OF HYDROCARBONS (Continued)

| | FORMULA | MOLEC. WT. | BOILING POINT °F | MELTING POINT °F | DENSITY | | | CRITICAL CONSTANTS | | | ACENTRIC FACTOR | HEAT OF VAPORIZ. @ B.P. Btu/lb | HEAT OF COMBUSTION @ 60°F Btu/lb | |
					°API	Sp Gr 60/60	Lb/Gal[a]	t °F	P Atm	D g/ml			GROSS	NET
PARAFFINS (continued)														
n-Octane	C_8H_{18}	114.2	285.2	- 70.2	68.6	0.707	5.89	564.1	24.5	0.232	0.394	129.5	20604	19099
2-Methylheptane	C_8H_{18}	114.2	243.8	-164.3	70.1	0.702	5.85	547.6	24.5	0.234	0.373	127.2	20585	19079
3-Ethylhexane	C_8H_{18}	114.2	245.4	-	65.6	0.718	5.98	558.1	25.7	0.251	0.361	126.5	20601	19097
2,5-Dimethylhexane	C_8H_{18}	114.2	228.4	-132.2	71.2	0.698	5.81	530.3	24.5	0.237	0.360	122.8	20564	19059
2,2,4-Trimethylpentane (Isooctane)	C_8H_{18}	114.2	210.6	-161.3	71.8	0.696	5.80	519.3	25.3	0.244	0.305	116.7	20569	19064
n-Nonane	C_9H_{20}	128.3	303.4	- 64.3	64.5	0.722	6.01	610.5	22.6	0.234	0.44	123.8	20544	19056
2-Methyloctane	C_9H_{20}	128.3	289.9	-112.7	65.7	0.717	5.98	596.2	22.6	0.237	0.42	122.9	20525	19036
3-Methyloctane	C_9H_{20}	128.3	291.5	-161.7	63.7	0.725	6.04	602.3	23.1	0.243	0.41	123.3	20532	19044
2,2-Dimethylheptane	C_9H_{20}	128.3	270.8	-171.4	66.5	0.715	5.95	580.0	22.9	0.244	0.38	116.6	20499	19011
2,6-Dimethylheptane	C_9H_{20}	128.3	275.4	-153.2	66.9	0.713	5.94	580.5	22.7	0.240	0.40	119.1	20507	19019
3,3-Diethylpentane	C_9H_{20}	128.3	295.1	- 27.6	55.3	0.757	6.31	599.5	24.0	0.253	0.47	118.4	20514	19026
n-Decane	$C_{10}H_{22}$	142.3	345.4	- 21.4	61.3	0.734	6.11	651.9	20.8	0.236	0.48	118.7	20493	19017
3-Methylnonane	$C_{10}H_{22}$	142.3	334.0	-120.6	60.4	0.737	6.14	644.5	21.1	0.244	0.45	118.2	20483	19007
5-Methylnonane	$C_{10}H_{22}$	142.3	329.2	-125.9	60.6	0.737	6.14	637.7	21.1	0.248	0.45	116.9	20485	19009
2,2-Dimethyloctane	$C_{10}H_{22}$	142.3	311.0	-	62.7	0.728	6.07	624.0	21.0	0.245	0.40	133.9	20451	18974
2,3-Dimethyloctane	$C_{10}H_{22}$	142.3	326.8	-	59.3	0.742	6.18	644.1	21.6	0.251	0.42	115.4	20479	19004
2,4-Dimethyloctane	$C_{10}H_{22}$	142.3	307.0	-	62.1	0.731	6.09	619.2	21.1	0.251	0.40	114.3	20469	18993
2,5-Dimethyloctane	$C_{10}H_{22}$	142.3	316.0	-	59.7	0.740	6.16	625.7	21.2	0.250	0.42	114.3	20468	18994
2,7-Dimethyloctane	$C_{10}H_{22}$	142.3	319.8	- 65.0	62.7	0.728	6.07	625.6	20.7	0.241	0.41	115.6	20460	18986
n-Undecane	$C_{11}H_{24}$	156.3	384.6	- 14.1	58.7	0.744	6.20	690.0	19.4	0.240	0.51	114.2	20443	18990
n-Dodecane	$C_{12}H_{26}$	170.3	421.3	14.7	56.4	0.753	6.27	725.2	18.0	0.240	0.55	110.2	20410	18966

a. In air, @ 60°F.

PHYSICAL CONSTANTS OF HYDROCARBONS (Continued)

FORMULA	MOLEC. WT.	BOILING POINT °F	MELTING POINT °F	°API	DENSITY Sp Gr 60/60	Lb/Gal^a	CRITICAL CONSTANTS t °F	P Atm	D g/ml	ACENTRIC FACTOR	HEAT OF VAPORIZ. @ B.P. Btu/lb	HEAT OF COMBUSTION @ 60°F Btu/lb GROSS	NET
OLEFINS													
Ethylene C_2H_4	28.1	-154.7	-272.5	273^b	0.35^b	2.91^b	48.6	50.0	0.227	0.085	207.6	21637^c	20276^c
Propylene C_3H_6	42.1	-53.9	-301.5	140	0.522	4.35	197.2	45.6	0.233	0.147	188.2	21043^c	19681^c
1-Butene C_4H_8	56.1	20.7	-301.6	104	0.601	5.01	295.6	39.7	0.233	0.197	167.9	20834^c	19473^c
cis-2-Butene C_4H_8	56.1	38.7	-218.1	94.2	0.627	5.22	324.3	41.5	0.240	0.208	178.9	20782^c	19420^c
trans-2-Butene C_4H_8	56.1	33.6	-158.0	100	0.610	5.08	311.8	40.5	0.236	0.212	174.4	20748^c	19387^c
2-Methylpropene (Isobutylene) C_4H_8	56.1	19.6	-220.6	104	0.600	5.00	292.5	39.5	0.234	0.202	169.5	20705^c	19344^c
1-Pentene C_5H_{10}	70.1	85.9	-265.4	87.5	0.646	5.38	376.9	35	0.240	0.236	154.5	20546	19185
cis-2-Pentene C_5H_{10}	70.1	98.5	-240.5	82.6	0.661	5.50	397.	35	0.23	0.27	160.1	20499	19138
trans-2-Pentene C_5H_{10}	70.1	97.4	-220.5	85.2	0.653	5.44	396.	35	0.23	0.26	159.8	20472	19090
2-Methyl-1-butene C_5H_{10}	70.1	88.1	-215.6	84.2	0.656	5.46	378.	35	0.23	0.25	156.3	20451	19090
3-Methyl-1-butene C_5H_{10}	70.1	68.1	-271.3	92.0	0.633	5.27	351.	34	0.24	0.21	147.5	20505	19144
2-Methyl-2-butene C_5H_{10}	70.1	101.4	-208.8	80.3	0.668	5.56	387.	34	0.22	0.28	161.3	20404	19043
1-Hexene C_6H_{12}	84.2	146.3	-219.7	77.2	0.678	5.65	447.5	31.0	0.23	0.28	144.4	20465	19104
cis-2-Hexene C_6H_{12}	84.2	156.0	-222.1	73.0	0.692	5.76	458^d	32^d	0.24^d	0.32	148.8	20406	19045
trans-2-Hexene C_6H_{12}	84.2	154.2	-207.4	75.7	0.683	5.69	456^d	32^d	0.24^d	0.31	147.7	20397	19036
cis-3-Hexene C_6H_{12}	84.2	151.6	-216.1	75.1	0.685	5.70	453^d	32^d	0.24^d	0.31	146.6	20431	19070
trans-3-Hexene C_6H_{12}	84.2	152.8	-172.2	76.0	0.682	5.68	454^d	32^d	0.24^d	0.31	147.9	20395	19034
1-Heptene C_7H_{14}	98.2	200.6	-182.0	70.1	0.702	5.85	507.3	28.0	0.22	0.332	136.2	20408	19047
1-Octene C_8H_{16}	112.2	250.3	-151.1	65.3	0.719	5.99	560.1	26.0	0.22	0.381	129.3	20366	19005
DIOLEFINS													
Propadiene C_3H_4	40.1	-30.1	-213.3	106	0.595	4.96	248	54^d	0.25^d	0.15	218.0	20872^c	19922^c
1,2-Butadiene C_4H_6	54.1	51.5	-213.2	83.5	0.658	5.48	339^d	44^d	0.25^d	0.25	181.0	20621^c	19568^c

a. In air, @ 60°F.   b. Hypothetical (extrapolated) value.
c. Heat of combustion as a gas, otherwise as a liquid.   d. Estimated.

PHYSICAL CONSTANTS OF HYDROCARBONS (Continued)

FORMULA	MOLEC. WT.	BOILING POINT °F	MELTING POINT °F	DENSITY			CRITICAL CONSTANTS			ACENTRIC FACTOR	HEAT OF VAPORIZ. @ B.P. Btu/lb	HEAT OF COMBUSTION @ 60°F Btu/lb	
				°API	Sp Gr 60/60	Lb/Gal[a]	t °F	P Atm	D g/ml			GROSS	NET
DIOLEFINS (continued)													
1,3-Butadiene	C4H6	54.1	24.1	-164.0	94.2	0.627	5.22	306	42.7	0.245	0.199	174.0	20208[c]	19152[c]
1,2-Pentadiene	C5H8	68.1	112.7	-215.1	71.2	0.698	5.81	446[d]	40[d]	0.25[d]	0.18	160.0	20347	19227
1,cis-3-Pentadiene	C5H8	68.1	111.3	-221.5	71.8	0.696	5.80	438[d]	40[d]	0.25[d]	0.20	160.0	19984	18864
1,trans-3-Pentadiene	C5H8	68.1	107.7	-125.4	76.3	0.681	5.67	434[d]	39[d]	0.25[d]	0.19	158.0	19944	18823
1,4-Pentadiene	C5H8	68.1	78.7	-234.9	81.0	0.666	5.55	400[d]	37[d]	0.25[d]	0.13	146.0	20151	19031
3-Methyl-1,2-butadiene	C5H8	68.1	105.5	-172.5	75.1	0.685	5.70	434[d]	39[d]	0.25[d]	0.17	156.0	20280	19158
2-Methyl-1,3-butadiene (Isoprene)	C5H8	68.1	93.3	-230.7	74.8	0.686	5.71	412[d]	38[d]	0.25[d]	0.18	153.0	19951	18830
1,5-Hexadiene	C6H10	82.1	139.0	-221.3	71.5	0.697	5.80	478[d]	34[d]	0.25[d]	0.16	134.0	20138	18960
2,3-Dimethyl-1,3-butadiene	C6H10	82.1	155.8	-104.9	62.1	0.731	6.09	489[d]	35[d]	0.25[d]	0.22	141.0	19890	18706
ACETYLENES													
Acetylene	C2H2	26.0	-119[e]	-114.0	209	0.416	3.46	95.3	60.6	0.231	0.207	-	21463[c]	20730[c]
Propyne (Methylacetylene)	C3H4	40.1	- 9.8	-152.9	94.9	0.625	5.20	264.6	55.5	0.245	0.226	175.0	20801[c]	19848[c]
1-Butyne (Ethylacetylene)	C4H6	54.1	46.5	-194.3	86.2	0.650	5.41	375.0	47[d]	0.24[d]	0.10	179.0	20643[c]	19587[c]
2-Butyne (Dimethylacetylene)	C4H6	54.1	80.6	- 26.0	71.5	0.697	5.80	419.	50[d]	0.24[d]	0.16	197.0	20281	19223
1-Pentyne (Propylacetylene)	C5H8	68.1	104.3	-158.3	72.1	0.695	5.79	428.	40[d]	0.25[d]	0.20	157.0	20374	19251
2-Pentyne	C5H8	68.1	132.9	-164.7	66.1	0.716	5.96	482[d]	42[d]	0.25[d]	0.18	167.0	20263	19140
3-Methyl-1-butyne	C5H8	68.1	79.4	-130.5	79.1	0.672	5.60	403[d]	40[d]	0.25[d]	0.14	-	20340	19217
1-Hexyne	C6H10	82.1	160.4	-205.4	64.8	0.721	6.00	506[d]	35[d]	0.25[d]	0.19	140.0	20351	19178
2-Hexyne	C6H10	82.1	184.1	-129.1	60.8	0.736	6.13	534[d]	34[d]	0.25[d]	0.21	-	-	-
3-Hexyne	C6H10	82.1	178.6	-153.6	63.1	0.727	6.05	538[d]	35[d]	0.25[d]	0.18	144.0	20259	19099

a. In air, @ 60°F.   c. Heat of combustion as a gas, otherwise as a liquid.
d. Estimated.   e. Sublimes.

PHYSICAL CONSTANTS OF HYDROCARBONS (Continued)

FORMULA	MOLEC. WT.	BOILING POINT °F	MELTING POINT °F	DENSITY			CRITICAL CONSTANTS			ACENTRIC FACTOR	HEAT OF VAPORIZ. @ B.P. Btu/lb	HEAT OF COMBUSTION @ 60°F Btu/lb	
				°API	Sp Gr 60/60	Lb/Gal[a]	t °F	P Atm	D g/ml			GROSS	NET
OLEFIN-ACETYLENES													
3-Butene-1-yne (Vinylacetylene)	C_4H_4	52.1	42.0	-	73.9	0.689	5.74	361[d]	48[d]	0.26[d]	0.13	-	-	-
1-Pentene-3-yne	C_5H_6	66.1	138.6	-	58.7	0.744	6.20	492[d]	41[d]	0.26[d]	0.17	-	-	-
1-Pentene-4-yne (Allylacetylene)	C_5H_6	66.1	107.0	-	49.4	0.782	6.51	441[d]	41[d]	0.26[d]	0.17	-	-	-
2-Methyl-1-butene-3-yne	C_5H_6	66.1	90.0	-	78.1	0.675	5.62	414[d]	41[d]	0.26[d]	0.17	-	-	-
AROMATICS													
Benzene	C_6H_6	78.1	176.2	42.0	28.4	0.885	7.37	552.1	48.3	0.302	0.220	169.3	17991	17258
Toluene	C_7H_8	92.1	231.1	-138.9	30.8	0.872	7.26	605.4	40.5	0.292	0.267	154.8	18251	17422
o-Xylene	C_8H_{10}	106.2	291.9	-13.3	28.4	0.885	7.37	674.8	36.8	0.288	0.297	149.1	18445	17546
m-Xylene	C_8H_{10}	106.2	282.4	-54.1	31.3	0.869	7.24	650.9	35.0	0.282	0.318	147.2	18441	17543
p-Xylene	C_8H_{10}	106.2	281.0	55.9	31.9	0.866	7.21	649.4	34.7	0.280	0.306	144.5	18445	17547
Ethylbenzene	C_8H_{10}	106.2	277.1	-138.9	30.8	0.872	7.26	651.1	35.6	0.284	0.307	144.0	18493	17595
1,2,3-Trimethylbenzene	C_9H_{12}	120.2	349.0	-13.7	25.9	0.899	7.49	736.3	34.1	0.313	0.315	143.2	18601	17649
1,2,4-Trimethylbenzene (Pseudocumene)	C_9H_{12}	120.2	336.8	-46.8	29.3	0.880	7.33	708.6	31.9	0.313	0.345	140.4	18590	17637
1,3,5-Trimethylbenzene (Mesitylene)	C_9H_{12}	120.2	328.5	-48.5	31.1	0.870	7.25	687.4	30.9	0.313	0.383	139.6	18584	17631
n-Propylbenzene	C_9H_{12}	120.2	318.6	-147.1	31.7	0.867	7.22	689.3	31.6	0.273	0.346	136.8	18674	17721
Isopropylbenzene (Cumene)	C_9H_{12}	120.2	306.3	-140.8	31.9	0.866	7.21	676.2	31.7	0.285	0.293	134.3	18665	17711
1-Methyl-2-ethylbenzene	C_9H_{12}	120.2	329.3	-113.4	28.4	0.885	7.27	712.	30.	0.26	0.31	139.0	18545	17692
1-Methyl-3-ethylbenzene	C_9H_{12}	120.2	322.4	-140.0	31.3	0.869	7.24	687.	28.	0.25	0.38	137.8	18637	17684
1-Methyl-4-ethylbenzene	C_9H_{12}	120.2	323.6	-80.2	31.9	0.866	7.21	693.	29.	0.26	0.39	137.4	18634	17680
1,2,4,5-Tetramethylbenzene (Durene)	$C_{10}H_{14}$	134.2	386.2	174.6	27.1[g]	0.892[g]	7.43[g]	756.	29.	0.28	0.43	145.8	-	-
Naphthalene	$C_{10}H_8$	128.2	424.3	176.5	7.98	1.015[g]	8.45[g]	887.5	40.0	0.31	0.30	-	17302[f]	16721[f]

a. In air, @ 60°F.   d. Estimated.
f. Heat of combustion as a solid.   g. For the supercooled liquid below normal melting point.

PHYSICAL CONSTANTS OF HYDROCARBONS (Continued)

FORMULA	MOLEC. WT	BOILING POINT °F	MELTING POINT °F	DENSITY			CRITICAL CONSTANTS			ACENTRIC FACTOR	HEAT OF VAPORIZ. @ B.P. Btu/lb	HEAT OF COMBUSTION @ 60°F Btu/lb	
				°API	Sp Gr 60/60	Lb/Gal[a]	t °F	P Atm	D g/ml			GROSS	NET
AROMATICS (continued)													
1,2,3,4-Tetrahydronaphthalene (Tetralin)	$C_{10}H_{12}$	132.2	405.6	-32.4	13.6	0.975	8.12	830[d]	33[d]	0.30[d]	0.33	-	-	-
cis-Decahydronaphthalene (cis-Decalin)	$C_{10}H_{18}$	138.2	384.2	-45.4	25.5	0.901	7.50	804.2	29[d]	0.28[d]	0.33	-	19567	18324
trans-Decahydronaphthalene (trans-Decalin)	$C_{10}H_{18}$	138.2	369.1	-22.7	30.4	0.874	7.28	776.8	29[d]	0.28[d]	0.33	-	19532	18289
NAPHTHENES													
Cyclopropane	C_3H_6	42.1	-27.0	-197.4	120	0.564	4.69	256.4	54.2	0.246	0.135	204.8	21378[c]	20017[c]
Cyclobutane	C_4H_8	56.1	54.5	-131.3	70.7	0.700	5.82	368.2	49.2	0.267	0.189	185.3	21038[c]	19673[c]
Cyclopentane	C_5H_{10}	70.1	120.7	-136.9	56.9	0.751	6.25	461.3	44.5	0.270	0.203	167.3	20188	18827
Methylcyclopentane	C_6H_{12}	84.1	161.3	-224.4	56.2	0.754	6.28	499.2	37.4	0.264	0.238	147.8	20130	18769
Cyclohexane	C_6H_{12}	84.1	177.3	43.8	49.2	0.783	6.52	536.5	40.2	0.273	0.213	153.0	20034	18676
Methylcyclohexane	C_7H_{14}	98.2	213.7	-195.9	51.3	0.774	6.45	570.2	34.3	0.267	0.242	136.3	20000	18639
Ethylcyclopentane	C_7H_{14}	98.2	218.2	-217.2	52.0	0.771	6.42	565.3	33.5	0.262	0.284	141.4	20120	18759
1,1-Dimethylcyclopentane	C_7H_{14}	98.2	190.1	-93.6	54.9	0.759	6.32	530[d]	35[d]	0.28[d]	0.27	132.6	20082	18721
1,cis-2-Dimethylcyclopentane	C_7H_{14}	98.2	211.2	-65.0	50.6	0.777	6.47	557[d]	34[d]	0.27[d]	0.27	138.8	20113	18752
1,trans-2-Dimethylcyclopentane	C_7H_{14}	98.2	197.4	-179.6	55.7	0.756	6.30	539[d]	34[d]	0.27[d]	0.26	135.1	20086	18724
1,cis-3-Dimethylcyclopentane	C_7H_{14}	98.2	195.4	-208.7	57.2	0.750	6.25	539[d]	35[d]	0.28[d]	0.26	134.3	20091	18730
1,trans-3-Dimethylcyclopentane	C_7H_{14}	98.2	197.1	-209.2	56.2	0.754	6.28	539[d]	34[d]	0.27[d]	0.26	134.9	20100	18739
Cyclopentene	C_5H_8	68.1	111.6	-211.2	45.6	0.799	6.65	451.2	47.3	0.275	0.168	-	19672	18551
Methylcyclopentadiene	C_6H_8	80.1	165.0	-	41.1	0.820	6.83	521[d]	42[d]	0.28[d]	0.22	-	-	-
1,5-Cyclooctadiene	C_8H_{12}	108.2	301.0	-69.5	28.0	0.887	7.39	704[d]	38[d]	0.30[d]	0.28	-	-	-

a. In air, @ 60°F. c. Heat of combustion as a gas, otherwise as a liquid.
d. Estimated.

185

PHYSICAL CONSTANTS OF ORGANIC COMPOUNDS

| FORMULA | MOLEC. WT. | BOILING POINT °F | MELTING POINT °F | DENSITY | | CRITICAL CONSTANTS | | | ACENTRIC FACTOR | HEAT OF VAPORIZ @ B.P. Btu/lb | HEAT OF COMBUSTION @ 60°F Btu/lb | |
				Sp Gr 60/60	Lb/Gal[a]	t °F	P Atm	D g/ml			GROSS	NET
ALCOHOLS												
Methanol (Methyl alcohol) CH$_3$OH	32.0	148.5	-143.8	0.796	6.64	463.0	79.9	0.272	0.569	464	9770	8570
Ethanol (Ethyl alcohol) CH$_3$CH$_2$OH	46.1	172.9	-173.4	0.794	6.62	469.6	63.0	0.276	0.635	361	12750	11510
1-Propanol (n-Propyl alcohol) CH$_3$CH$_2$CH$_2$OH	60.1	207.0	-195.2	0.808	6.74	506.4	51.0	0.275	0.621	299	14460	13190
2-Propanol (Isopropyl alcohol) (CH$_3$)$_2$CHOH	60.1	180.2	-127.3	0.790	6.59	455.3	47.0	0.273	0.666	285	14350	13080
1-Butanol (n-Butyl alcohol) CH$_3$CH$_2$CH$_2$CH$_2$OH	74.1	243.8	-128.7	0.814	6.78	553.6	43.6	0.270	0.590	250	15530	14250
2-Butanol (sec-Butyl alcohol) CH$_3$CH$_2$CH(OH)CH$_3$	74.1	211.3	-174.5	0.812	6.77	505.0	41.4	0.276	0.578	237	15490	14200
2-Methyl-1-propanol (Isobutyl alcohol) (CH$_3$)$_2$CHCH$_2$OH	74.1	225.8	-162.0	0.806	6.72	526.2	42.4	0.273	0.587	244	15450	14160
2-Methyl-2-propanol (tert-Butyl alcohol) (CH$_3$)$_3$COH	74.1	180.4	78.2	0.792	6.60	451.4	39.2	0.270	0.615	226	15350	14060
1-Pentanol CH$_3$(CH$_2$)$_3$CH$_2$OH	88.2	280.0	-108.8	0.819	6.83	595.0	37[d]	0.27[d]	0.58	216	16240	14940
2-Pentanol CH$_3$(CH$_2$)$_2$CH(OH)CH$_3$	88.2	246.2	-	0.814	6.78	560[d]	35[d]	0.27[d]	0.48	210	16170	14870
3-Pentanol (CH$_3$CH$_2$)$_2$CHOH	88.2	239.5	-	0.825	6.88	550[d]	35[d]	0.27[d]	0.48	206	16170	14870
2-Methyl-1-butanol CH$_3$CH$_2$CH(CH$_3$)CH$_2$OH	88.2	263.7	-	0.823	6.86	593[d]	38[d]	0.27[d]	0.47	214	16220	14920
2-Methyl-2-butanol CH$_3$CH$_2$C(OH)(CH$_3$)$_2$	88.2	215.6	16.2	0.814	6.79	522.	34[d]	0.27[d]	0.44	196	16110	14810
3-Methyl-1-butanol (CH$_3$)$_2$CHCH$_2$CH$_2$OH	88.2	266.9	-	0.814	6.79	583.3	38[d]	0.27[d]	0.50	215	16220	14920
3-Methyl-2-butanol (CH$_3$)$_2$CHCH(OH)CH$_3$	88.2	232.7	-	0.822	6.86	550	36[d]	0.27[d]	0.44	202	16170	14870
2,2-Dimethyl-1-propanol (CH$_3$)$_3$CCH$_2$OH	88.2	235.6	125.6	0.823	6.86	555[d]	38[d]	0.27[d]	0.46	213	-	-
1-Octanol CH$_3$(CH$_2$)$_6$CH$_2$OH	130.2	383.4	4.1	0.830	6.92	725.0	29[d]	0.27[d]	0.56	155	17490	16170
2-Ethyl-1-hexanol CH$_3$(CH$_2$)$_3$CH(C$_2$H$_5$)CH$_2$OH	130.2	364.3	-94.0	0.837	6.98	700[d]	30[d]	0.27[d]	0.56	170[d]	17470	16150
1-Tridecanol CH$_3$(CH$_2$)$_{11}$CH$_2$OH	200.4	525.2	87.1	0.835	6.97	875[d]	21[d]	0.26[d]	0.6[d]	130[d]	-	-
1-Hexadecanol (Cetyl alcohol) CH$_3$(CH$_2$)$_{14}$CH$_2$OH	242.4	593.6	120.6	0.842	7.02	920[d]	17[d]	0.26[d]	0.7[d]	115[d]	18580	17240

a. In air, @ 60°F.   d. Estimated.

PHYSICAL CONSTANTS OF ORGANIC COMPOUNDS (Continued)

FORMULA		MOLEC. WT.	BOILING POINT °F	MELTING POINT °F	DENSITY		CRITICAL CONSTANTS			ACENTRIC FACTOR	HEAT OF VAPORIZ. @ B.P. Btu/lb	HEAT OF COMBUSTION @ 60°F Btu/lb	
					Sp Gr 60/60	Lb/Gal[a]	t °F	P Atm	D g/ml			GROSS	NET
GLYCOLS AND GLYCEROL													
1,2-Ethanediol (Ethylene glycol)	CH_2OHCH_2OH	62.1	387.1	9.6	1.118	9.32	700^d	75^d	0.33^d	-	380^d	8240	7320
1,2-Propanediol (Propylene glycol)	$CH_3CH(OH)CH_2OH$	76.1	369.7	-76^h	1.040	8.67	670^d	60^d	0.32^d	-	300^d	10310	9300
1,3-Propanediol (Trimethylene glycol)	$CH_2OHCH_2CH_2OH$	76.1	417.9	-16.1	1.057	8.82	725^d	60^d	0.32^d	-	320^d	-	-
1,2,3-Propanetriol (Glycerol)	$CH_2(OH)CH(OH)CH_2OH$	92.1	554.0	65	1.265	10.53	850^d	65^d	0.36^d	-	310^d	10850	10020
ETHERS													
Dimethyl ether	CH_3OCH_3	46.1	-12.7	-222.7	0.676	5.64	260.4	53.0	0.242	0.189	201	13630^c	12390^c
Diethyl ether	$CH_3CH_2OCH_2CH_3$	74.1	94.2	-177.3	0.719	6.00	380.4	35.9	0.265	0.281	155	15810	14530
Di-n-propyl ether	$CH_3(CH_2)_2O(CH_2)_2CH_3$	102.2	193.4	-189.8	0.752	6.27	490^d	29^d	0.26^d	0.38	133	16050	14810
Diisopropyl ether	$(CH_3)_2CHOCH(CH_3)_2$	102.2	154.9	-121.9	0.729	6.08	440.4	28.4	0.265	0.341	123	15960	14720
Di-n-butyl ether	$CH_3(CH_2)_3O(CH_2)_3CH_3$	130.2	288.0	-144.0	0.773	6.44	585^d	25^d	0.27^d	0.51	122	17650	16330
Di-sec-butyl ether	$[CH_3CH_2CH(CH_3)]_2O$	130.2	250.0	-	0.760	6.34	546^d	25^d	0.27^d	0.43^d	114^d	17570	16260
ALDEHYDES													
Methanal (Formaldehyde)	HCHO	30.0	-2.4	-133.6	-	-	-	-	-	-	331^d	8180^c	7540^c
Ethanal (Acetaldehyde)	CH_3CHO	44.0	68.7	-189.4	0.785	6.54	370.	55^d	0.26^d	0.31	262^d	11380	10510
Propanal (Propionaldehyde)	CH_3CH_2CHO	58.1	118.4	-112.0	0.803	6.69	475^d	45^d	0.26^d	0.34	217^d	13460	12470
n-Butanal (Butyraldehyde)	$CH_3CH_2CH_2CHO$	72.1	166.6	-141.5	0.807	6.73	485^d	40^d	0.26^d	0.35	191^d	14790	13730
2-Methylpropanal (Isobutyraldehyde)	$(CH_3)_2CHCHO$	72.1	147.4	-85.0	0.795	6.63	465^d	40^d	0.26^d	0.35	185^d	14770	13660
n-Octanal (Caprylaldehyde)	$CH_3(CH_2)_5CH_2CHO$	128.2	345.2	-16.6	0.825	6.88	655^d	25^d	0.26^d	0.56	136^d	-	-

a. In air, @ 60°F.   c. Heat of combustion as a gas, otherwise as a liquid.
d. Estimated.   h. Sets to a glass below this temperature.

PHYSICAL CONSTANTS OF ORGANIC COMPOUNDS (Continued)

	FORMULA	MOLEC. WT.	BOILING POINT °F	MELTING POINT °F	DENSITY		CRITICAL CONSTANTS			ACENTRIC FACTOR	HEAT OF VAPORIZ. @ B.P. Btu/lb	HEAT OF COMBUSTION @ 60°F Btu/lb	
					Sp Gr 60/60	Lb/Gal[a]	t °F	P Atm	D g/ml			GROSS	NET
KETONES													
2-Propanone (Acetone)	CH_3COCH_3	58.1	133.3	-138.5	0.796	6.63	455.0	46.4	0.278	0.310	220	13250	12270
2-Butanone (Methyl ethyl ketone)	$CH_3COCH_2CH_3$	72.1	175.4	-124.4	0.810	6.76	504.5	41.0	0.270	0.359	190	14580	13520
2-Pentanone (Methyl n-propyl ketone)	$CH_3COCH_2CH_2CH_3$	86.1	216.2	-106.4	0.812	6.77	555.4	38.4	0.286	0.34	171[d]	15480	14380
3-Pentanone (Diethyl ketone)	$(CH_3CH_2)_2CO$	86.1	215.6	-38.2	0.819	6.83	550.0	36.9	0.256	0.34	171[d]	15490	14390
3-Methyl-2-butanone (Methyl isopropyl ketone)	$CH_3COCH(CH_3)_2$	86.1	201.9	-133.6	0.808	6.74	536.4	38.0	0.278	0.35	167[d]	15470	14370
4-Methyl-2-butanone (Methyl isobutyl ketone)	$CH_3COCH_2CH(CH_3)_2$	100.2	241.7	-119.2	0.806	6.72	568.	32.3	0.28[d]	0.39	153[d]	-	-
4-Methyl-3-pentene-2-one (Mesityl oxide)	$(CH_3)_2C:CHCOCH_3$	98.1	266.0	-63.2	0.860	7.20	620[d]	35[d]	0.29[d]	0.36	162[d]	-	-
ACIDS													
Methanoic acid (Formic acid)	HCOOH	46.0	213.0	47.1	1.227	10.23	-	-	-	-	350[d]	2370	1960
Ethanoic acid (Acetic acid)	CH_3COOH	60.1	244.2	62.0	1.055	8.80	610.3	57.1	0.351	0.446	280[d]	6270	5630
Propanoic acid (Propionic acid)	CH_3CH_2COOH	74.1	285.5	-5.3	0.999	8.33	662[d]	53[d]	0.32[d]	0.54	240[d]	8870	8090
n-Butanoic acid (Butyric acid)	$CH_3CH_2CH_2COOH$	88.1	325.9	22.6	0.964	8.03	671[d]	52[d]	0.30[d]	0.67	210[d]	12680	11650
2-Methylpropanoic acid (Isobutyric acid)	$(CH_3)_2CHCOOH$	88.1	310.5	-50.8	0.954	7.95	637[d]	40[d]	0.30[d]	0.67	210[d]	-	-

a. In air, @ 60°F.   d. Estimated.

PHYSICAL CONSTANTS OF MISCELLANEOUS MATERIALS

	FORMULA	MOLEC. WT.	BOILING POINT °F	MELTING POINT °F	DENSITY Sp Gr 60/60	DENSITY Lb/Gal	CRITICAL CONSTANTS t °F	CRITICAL CONSTANTS P Atm	CRITICAL CONSTANTS D g/ml	ACENTRIC FACTOR	HEAT OF VAPORIZ. @ B.P. Btu/lb	HEAT OF COMBUSTION @ 60 F Btu/lb GROSS	HEAT OF COMBUSTION @ 60 F Btu/lb NET
Acetonitrile	CH_3CN	41.1	178.9	- 50.3	0.787	6.55	526.	47.7	0.237	0.323	329.5	-	-
Ammonia	NH_3	17.0	- 28.1	-107.9	0.617	5.15	270.5	112.5	0.235	0.257	590.9	-	-
Argon	Ar	40.0	-302.5	-308.9	1.392^i	11.62	-188.5	48.0	0.531	0	70.0	-	-
Carbon dioxide	CO_2	44.0	-109.2^e	- 69.9	0.820	6.84	87.8	72.8	0.468	0.225	149.6^j	-	-
Carbon monoxide	CO	28.0	-312.7	-337.0	0.789^i	6.58	-220.4	34.5	0.301	0.049	92.8	4345^c	4345^c
Carbonyl sulfide	COS	60.1	- 58.4	-217.8	1.005	8.39	216.	58.	0.44	0.13	-	-	-
Chlorine	Cl_2	70.9	- 29.3	-149.8	1.427	11.88	291.2	76.1	0.573	0.073	123.8	-	-
Diethanolamine	$HN(CH_2CH_2OH)_2$	105.1	515.1	77.2	1.098	9.14	828^d	32^d	0.28^d	1.04	267.0	-	-
Dinitrogen tetroxide	N_2O_4	92.0	69.8	11.8	1.457	12.16	316.8	100.	0.55	0.852	178.2	-	-
Ethyl chloride	CH_3CH_2Cl	64.5	54.3	-216.9	0.903	7.54	369.0	52.	0.33	0.192	164.7	-	-
Ethyl mercaptan	CH_3CH_2SH	62.1	95.0	-234.2	0.845	7.05	439.	54.2	0.300	0.187	185.5	-	-
Helium	He	4.0	-452.1	-455.8	0.125^i	1.04	-450.3	2.3	0.069	-0.387	8.7	-	-
Hydrogen	H_2	2.0	-423.0	-434.6	0.0711^i	0.59	-399.8	12.8	0.031	-0.223	194.4	61100^c	51600^c
Hydrogen chloride	HCl	36.5	-121.0	-173.5	0.827	6.90	124.7	82.0	0.45	0.125	190.6	-	-
Hydrogen fluoride	HF	20.0	67.1	-118.1	0.986	8.23	370.0	64.0	0.29	0.374	175.5	-	-
Hydrogen sulfide	H_2S	34.1	- 76.6	-122.0	0.786	6.55	212.0	88.2	0.346	0.100	235.6	-	-
Methyl chloride	CH_3Cl	50.5	- 11.6	-143.5	0.929	7.75	289.6	65.9	0.363	0.155	182.3	-	-
Methyl mercaptan	CH_3SH	48.0	42.7	-189.3	0.873	7.28	386.2	71.4	0.332	0.151	220.2	-	-
Monoethanolamine	$H_2NCH_2CH_2OH$	61.1	338.6	50.5	1.015	8.45	646^d	44^d	0.28^d	0.84	357.9	-	-
Nitric oxide	NO	30.0	-241.1	-262.5	1.27^i	10.60	-135.2	64.6	0.52	0.489	197.6	-	-
Nitrogen	N_2	28.0	-320.4	-346.0	0.808^i	6.74	-232.6	33.5	0.311	0.040	85.7	-	-
Nitrous oxide	N_2O	44.0	-127.2	-131.5	0.788	6.58	97.6	71.5	0.452	0.162	161.8	-	-

a. In air, @ 60°F.   c. Heat of combustion as a gas, otherwise as a liquid.
d. Estimated.   e. Sublimes.   j. At triple point.
i. At normal boiling point.

PHYSICAL CONSTANTS OF MISCELLANEOUS MATERIALS (Continued)

	FORMULA	MOLEC. WT.	BOILING POINT °F	MELTING POINT °F	DENSITY Sp Gr 60/60	DENSITY Lb/Gal[a]	CRITICAL CONSTANTS t °F	CRITICAL CONSTANTS P Atm	CRITICAL CONSTANTS D g/ml	ACENTRIC FACTOR	HEAT OF VAPORIZ. @ B.P. Btu/lb	HEAT OF COMBUSTION @ 60°F Btu/lb GROSS	HEAT OF COMBUSTION @ 60°F Btu/lb NET
Oxygen	O_2	32.0	-297.4	-361.9	1.142[i]	9.53	-181.5	49.8	0.436	0.021	91.7	-	-
Phenol	C_6H_5OH	93.1	359.3	105.6	1.08[g]	9.00	790	60.5	0.356	0.443	211.1	-	-
Sulfur dioxide	SO_2	64.1	14.0	-103.9	1.394	11.63	315.7	77.8	0.525	0.256	170.8	-	-
Sulfur trioxide	SO_3	80.1	112.6	62.3	1.925	16.06	423.9	81.	0.633	0.392	233.5	-	-
Vinyl chloride	$CH_2:CHCl$	62.5	7.9	-244.8	0.969[i]	8.07	315[d]	55[d]	0.37[d]	0.098	143.0	-	-
Water	H_2O	18.0	212.0	32.0	1.000	8.328	705.5	218.3	0.315	0.344	970.6	-	-

a. In air, @ 60°F.
g. For the supercooled liquid below normal melting point.
d. estimated
i. At normal boiling point.

CONVERSION FACTORS

BASIC CONSTANTS

Absolute temperature of the "ice" point	$T = 273.160° K$
	$= 491.688° R$
Volume of an ideal gas at 60°F and 1 atm pressure	$V = 379.5$ cu ft/lb-mole
Pressure-volume product for one mole of a gas at 0°C and zero pressure	$(PV) \begin{smallmatrix} P = 0 \\ T = 0°C \end{smallmatrix} = 22.4140$ (liter) (atm)/(g-mole)
Gas Constant	$R = 0.0820544$ (liter) (atm)/(g-mole) (°K)
Standard gravity	$g_0 = 980.665$ cm/sec^2
	$= 32.174$ ft/sec^2
Standard atmosphere	$Atm = 14.6960$ lbs/sq in. abs
Calorie (thermochemical)	$Cal = 4.1840$ absolute joules
Natural logarithm	$\ln X$ or $\log_e X = 2.30259 \log_{10} X$
Base of natural logarithms	$e = 2.71828$

GAS CONSTANTS (R)

The gas constant R appears with various other numerical values. The value for use should be selected so that the units are consistent with those of the other terms with which R is to be used.

R	Units
10.731	(cu ft) (lbs per sq in. abs)/(lb-mole) (°R)
1,545.3	(cu ft) (lbs per sq ft abs)/(lb-mole) (°R)
0.7302	(cu ft) (atm)/lb-mole) (°R)
0.08205	(liter) (atm)/(g-mole) (°K)
82.06	(cu cm) (atm)/g-mole) (°K)
62.36	(liter) (mm of Hg) (g-mole) (°K)
62,363.	(cu cm) (mm of Hg)/(g-mole) (°K)
1.9872	(g-cal)/(g-mole) (°K)
1.9872	(Btu)/(lb-mole) (°R)
8.314	(joules)/(g-mole) (°K)

CONVERSION FACTORS (Continued)

LENGTH							
TO CONVERT FROM ↓	Multiply by factor below to obtain						
	Millimeters	Centimeters	Meters	Inches	Feet		
Millimeters	1	0.1	0.001	0.03937	0.003281		
Centimeters	10	1	0.01	0.3937	0.03281		
Meters	1000	100	1	39.37*	3.281		
Inches	25.40	2.540	0.02540	1	0.08333		
Feet	304.8	30.48	0.3048	12	1		
—							

1 statute mile (U.S.)   = 5,280 feet
1 nautical mile (U.S.) = 6,080.20 feet
1 kilometer   = 0.6214 miles (U.S.)
1 Ångstrom unit (Å)   = 10^{-8} centimeters
1 micron (μ)   = 10^{-6} meters

AREA						
TO CONVERT FROM ↓	Multiply by factor below to obtain					
	Square centimeters	Square meters	Square inches	Square feet		
Square centimeters	1	10^{-4}	0.15500	0.0010764		
Square meters	10,000	1	1550.0	10.764		
Square inches	6.452	6.452×10^{-4}	1	0.006944		
Square feet	929.0	0.09290	144	1		

1 acre   = 43,560 square feet
1 hectare = 10,000 square meters

VOLUME							
TO CONVERT FROM ↓	Multiply by factor below to obtain						
	Cubic inches	Cubic feet	U. S. gallons	Imperial gallons	Cubic centimeters	Liters	Barrels (42's)
Cubic inches	1	5.787×10^{-4}	0.004329	0.003605	16.387	0.016387	1.0307×10^{-4}
Cubic feet	1,728	1	7.481	6.229	28,317.	28.32	0.17811
U. S. gallons	231*	0.13368	1	0.8327	3,785.	3.785	0.02381
Imperial gallons	277.4	0.16054	1.2009	1	4,546.	4.546	0.02859
Cubic centimeters	0.06102	3.531×10^{-5}	2.642×10^{-4}	2.200×10^{-4}	1	1.0000×10^{-3}	6.290×10^{-6}
Liters	61.03	0.03532	0.2642	.2200	1,000.0	1	0.006290
Barrels (42's)	9,702	5.615	42	34.97	1.5898×10^5	158.98	1

1 milliliter   = 1.000028 cubic centimeters
1 cubic meter = 6.290 barrels (42's)

*Exact value by definition

CONVERSION FACTORS (Continued)

MASS

TO CONVERT FROM	Multiply by factor below to obtain						
	Grams	Kilograms	Pounds	Short tons	Long tons	Metric tons	
Grams	1	10^{-3}	0.002205	1.1023×10^{-6}	0.9842×10^{-6}	10^{-6}	
Kilograms	1,000	1	2.205	1.1023×10^{-3}	0.9842×10^{-3}	10^{-3}	
Pounds	453.6	0.4536	1	5.000×10^{-4}	4.464×10^{-4}	4.536×10^{-4}	
Short tons	9.072×10^5	907.2	2,000	1	0.8929	0.9072	
Long tons	10.160×10^5	1,016.0	2,240	1.1200	1	1.0160	
Metric tons	10^6	10^3	2,205.	1.1023	0.9842	1	

1 pound (avoirdupois) = 16 ounces (avoirdupois)
                = 7000 grains
1 ounce (avoirdupois) = 437.5 grains
1 pound (troy)   = 12 ounces (troy)
1 ounce (troy)   = 480 grains

DENSITY[a]

TO CONVERT FROM	Multiply by factor below to obtain						
	Grams/ milliliter	Pounds/ cubic foot	Pounds/gallon (U.S.)	Pounds/barrel (42's)			
Grams/milliliter	1	62.43	8.345	350.5			
Pounds/cubic foot	0.016019	1	0.13368	5.615			
Pounds/gallon (U.S.)	0.11983	7.481	1	42			
Pounds/barrel (42's)	0.002853	0.17811	0.02381	1			

(a) The values in this table are based on weight in vacuum (true mass); values for pounds/gallon taken from API gravity conversion tables are based on weight in air. For engineering design purposes, the difference is generally negligible.

PRESSURE

TO CONVERT FROM	Multiply by factor below to obtain						
	Pounds/ square inch	Pounds/ square foot	Atmospheres	Kilograms/ sq centimeter	Inches of mercury	Millimeters of mercury	Feet of water (60°F)
Pounds/square inch	1	144	0.06805	0.07031	2.036	51.71	2.309
Pounds/square foot	0.006944	1	4.725×10^{-4}	4.882×10^{-4}	0.014139	0.3591	0.016034
Atmospheres	14.696	2116.	1	1.0332	29.92	760	33.93
Kilograms/sq cm	14.223	2048.	0.9678	1	28.96	735.6	32.84
Inches of mercury	0.4912	70.73	0.03342	0.03453	1	25.40	1.1340
Millimeters of mercury	0.019337	2.785	0.0013158	0.0013595	0.03937	1	0.04465
Feet of water (60°F)	0.4331	62.37	0.02947	0.03045	0.8818	22.40	1

1 atmosphere = 1,013,250 dynes/square centimeter
            = 1.013250 bars
1 torr   = 1 millimeter of mercury (pressure)

TEMPERATURE

TO CONVERT FROM	Use formula below to obtain			
	°Centigrade (°C)	°Fahrenheit (°F)	°Rankine (°R)	°Kelvin (°K)
°Centigrade (°C)	–	1.8 (°C) + 32	1.8 (°C) + 491.69	°C + 273.16
°Fahrenheit (°F)	(°F – 32)/1.8	–	°F + 459.69	(°F + 459.69)/1.8
°Rankine (°R)	(°R – 491.69)/1.8	°R – 459.69	–	°R/1.8
°Kelvin (°K)	°K – 273.16	1.8 (°K) – 459.69	1.8 (°K)	–

CONVERSION FACTORS (Continued)

RATE OF FLOW

TO CONVERT FROM ↓	Multiply by factor below to obtain						
	Liters/second	Gallons (U.S.)/minute	Gallons (U.S.)/hour	Cubic feet/second	Cubic feet/minute	Barrels (42's)/day	Cubic meters/day
Liters/second	1	15.851	951.0	0.03531	2.119	543.5	86.40
Gallons (U.S.)/minute	0.06309	1	60	0.002228	0.13368	34.29	5.451
Gallons (U.S.)/hour	0.0010515	0.016667	1	3.713×10^{-5}	0.002228	0.5714	0.09085
Cubic feet/second	28.32	448.8	2.693×10^4	1	60	1.5388×10^4	2,447.
Cubic feet/minute	0.4719	7.481	448.8	0.016667	1	256.5	40.78
Barrels (42's)/day	0.0018400	0.02917	1.75	6.498×10^{-5}	3.899×10^{-3}	1	0.15898
Cubic meters/day	0.011574	0.18345	11.007	4.087×10^{-4}	0.02452	6.290	1

ENERGY, HEAT, AND WORK[a]

TO CONVERT FROM ↓	Multiply by factor below to obtain						
	Btu	Kilogram-calories	Foot-pounds	Horsepower-hours	Kilowatt-hours		
Btu	1	0.2522	778.2	3.930×10^{-4}	2.931×10^{-4}		
Kilogram-calories	3.966	1	3,086.	1.559×10^{-3}	1.162×10^{-3}		
Foot-pounds	1.2851×10^{-3}	3.240×10^{-4}	1	5.051×10^{-7}	3.766×10^{-7}		
Horsepower-hours	2,544.	641.6	198×10^4	1	0.7457		
Kilowatt-hours	3,412.	860.4	2.655×10^6	1.3410	1		

(a) Based on thermochemical calorie and I.T. (International Steam Tables) Btu.

1 calorie (thermochemical)	=	0.9994346 I.T. calorie
	=	4.1840 absolute joules
1 Btu (thermochemical)	=	0.9994346 I.T. Btu
1 I.T. Btu	=	1055.040 absolute joules
1 I.T. calorie/gram	=	1.8 I.T. Btu/pound

POWER[a]

TO CONVERT FROM ↓	Multiply by factor below to obtain						
	Btu/hour	Foot-pounds/minute	Foot-pounds/second	Horsepower	Kilowatts	Kilogram-cal/second	Tons of refrigeration
Btu/hour	1	12.969	0.2162	3.930×10^{-4}	2.931×10^{-4}	7.004×10^{-5}	8.333×10^{-5}
Foot-pounds/minute	0.07711	1	0.016667	3.030×10^{-5}	2.260×10^{-5}	5.401×10^{-6}	6.425×10^{-6}
Foot-pounds/second	4.626	60	1	1.8182×10^{-3}	1.3558×10^{-3}	3.241×10^{-4}	3.855×10^{-4}
Horsepower	2544.	33,000	550	1	0.7457	0.17823	0.2120
Kilowatts	3412.	44,254.	737.6	1.3410	1	0.2390	0.2843
Kilogram-cal/second	14,278.	185,160.	3086.	5.611	4.184	1	1.1897
Tons of refrigeration	12,000	155,630.	2594.	4.716	3.517	0.8405	1

(a) See Footnote (a) to table above.

THERMAL CONDUCTIVITY

Values of thermal conductivity are generally expressed as Btu per hr/(sq ft)(°F per ft). Other units may be converted as follows:

Btu per hr/(sq ft) (°F per inch)	x	0.0833	=	Btu per hr/(sq ft) (°F per ft)
gram-cal per sec/(sq cm) (°C per cm)	x	242.	=	Btu per hr/(sq ft) (°F per ft)
kilogram-cal per hr/(sq meter) (°C per meter)	x	0.672	=	Btu per hr/(sq ft) (°F per ft)
joules per sec (or watts)/(sq cm) (°C per cm)	x	57.8	=	Btu per hr/(sq ft) (°F per ft)

TEMPERATURE CONVERSION TABLES

−460° to 0°

TO °F	FROM	TO °C	TO °F	FROM	TO °C	TO °F	FROM	TO °C	TO °F	FROM	TO °C	TO °F	FROM	TO °C
				−400	−240.00		−300	−184.44	−328.0	−200	−128.89	−148.0	−100	−73.33
				−398	−238.89		−298	−183.33	−324.4	−198	−127.78	−144.4	− 98	−72.22
				−396	−237.78		−296	−182.22	−320.8	−196	−126.67	−140.8	− 96	−71.11
				−394	−236.67		−294	−181.11	−317.2	−194	−125.56	−137.2	− 94	−70.00
				−392	−235.56		−292	−180.00	−313.6	−192	−124.44	−133.6	− 92	−68.89
				−390	−234.44		−290	−178.89	−310.0	−190	−123.33	−130.0	− 90	−67.78
				−388	−233.33		−288	−177.78	−306.4	−188	−122.22	−126.4	− 88	−66.67
				−386	−232.22		−286	−176.67	−302.8	−186	−121.11	−122.8	− 86	−65.56
				−384	−231.11		−284	−175.56	−299.2	−184	−120.00	−119.2	− 84	−64.44
				−382	−230.00		−282	−174.44	−295.6	−182	−118.89	−115.6	− 82	−63.33
				−380	−228.89		−280	−173.33	−292.0	−180	−117.78	−112.0	− 80	−62.22
				−378	−227.78		−278	−172.22	−288.4	−178	−116.67	−108.4	− 78	−61.11
				−376	−226.67		−276	−171.11	−284.8	−176	−115.56	−104.8	− 76	−60.00
				−374	−225.56		−274	−170.00	−281.2	−174	−114.44	−101.2	− 74	−58.89
				−372	−224.44	−457.6	−272	−168.89	−277.6	−172	−113.33	− 97.6	− 72	−57.78
				−370	−223.33	−454.0	−270	−167.78	−274.0	−170	−112.22	− 94.0	− 70	−56.67
				−368	−222.22	−450.4	−268	−166.67	−270.4	−168	−111.11	− 90.4	− 68	−55.56
				−366	−221.11	−446.8	−266	−165.56	−266.8	−166	−110.00	− 86.8	− 66	−54.44
				−364	−220.00	−443.2	−264	−164.44	−263.2	−164	−108.89	− 83.2	− 64	−53.33
				−362	−218.89	−439.6	−262	−163.33	−259.6	−162	−107.78	− 79.6	− 62	−52.22
	−459.69	−273.16		−360	−217.78	−436.0	−260	−162.22	−256.0	−160	−106.67	− 76.0	− 60	−51.11
	−458	−272.22		−358	−216.67	−432.4	−258	−161.11	−252.4	−158	−105.56	− 72.4	− 58	−50.00
	−456	−271.11		−356	−215.56	−428.8	−256	−160.00	−248.8	−156	−104.44	− 68.8	− 56	−48.89
	−454	−270.00		−354	−214.44	−425.2	−254	−158.89	−245.2	−154	−103.33	− 65.2	− 54	−47.78
	−452	−268.89		−352	−213.33	−421.6	−252	−157.78	−241.6	−152	−102.22	− 61.6	− 52	−46.67
	−450	−267.78		−350	−212.22	−418.0	−250	−156.67	−238.0	−150	−101.11	− 58.0	− 50	−45.56
	−448	−266.67		−348	−211.11	−414.4	−248	−155.56	−234.4	−148	−100.00	− 54.4	− 48	−44.44
	−446	−265.56		−346	−210.00	−410.8	−246	−154.44	−230.8	−146	− 98.99	− 50.8	− 46	−43.33
	−444	−264.44		−344	−208.89	−407.2	−244	−153.33	−227.2	−144	− 97.78	− 47.2	− 44	−42.22
	−442	−263.33		−342	−207.78	−403.6	−242	−152.22	−223.6	−142	− 96.67	− 43.6	− 42	−41.11
	−440	−262.22		−340	−206.67	−400.0	−240	−151.11	−220.0	−140	− 95.56	− 40.0	− 40	−40.00
	−438	−261.11		−338	−205.56	−396.4	−238	−150.00	−216.4	−138	− 94.44	− 36.4	− 38	−38.89
	−436	−260.00		−336	−204.44	−392.8	−236	−148.89	−212.8	−136	− 93.33	− 32.8	− 36	−37.78
	−434	−258.89		−334	−203.33	−389.2	−234	−147.78	−209.2	−134	− 92.22	− 29.2	− 34	−36.67
	−432	−257.78		−332	−202.22	−385.6	−232	−146.67	−205.6	−132	− 91.11	− 25.6	− 32	−35.56
	−430	−256.67		−330	−201.11	−382.0	−230	−145.56	−202.0	−130	− 90.00	− 22.0	− 30	−34.44
	−428	−255.56		−328	−200.00	−378.4	−228	−144.44	−198.4	−128	− 88.89	− 18.4	− 28	−33.33
	−426	−254.44		−326	−198.89	−374.8	−226	−143.33	−194.8	−126	− 87.78	− 14.8	− 26	−32.22
	−424	−253.33		−324	−197.78	−371.2	−224	−142.22	−191.2	−124	− 86.67	− 11.2	− 24	−31.11
	−422	−252.22		−322	−196.67	−367.6	−222	−141.11	−187.6	−122	− 85.56	− 7.6	− 22	−30.00
	−420	−251.11		−320	−195.56	−364.0	−220	−140.00	−184.0	−120	− 84.44	− 4.0	− 20	−28.89
	−418	−250.00		−318	194.44	−360.4	−218	−138.89	−180.4	−118	− 83.33	− 0.4	− 18	−27.78
	−416	−248.89		−316	193.33	−356.8	−216	−137.78	−176.8	−116	− 82.22	+ 3.2	− 16	−26.67
	−414	−247.78		−314	192.22	−353.2	−214	−136.67	−173.2	−114	− 81.11	+ 6.8	− 14	−25.56
	−412	−246.67		−312	191.11	−349.6	−212	−135.56	−169.6	−112	− 80.00	+ 10.4	− 12	−24.44
	−410	−245.56		−310	190.00	−346.0	−210	−134.44	−166.0	−110	− 78.89	+ 14.0	− 10	−23.33
	−408	−244.44		−308	188.89	−342.4	−208	−133.33	−162.4	−108	− 77.78	+ 17.6	− 8	−22.22
	−406	−243.33		−306	187.78	−338.8	−206	−132.22	−158.8	−106	− 76.67	+ 21.2	− 6	−21.11
	−404	−242.22		−304	186.67	−335.2	−204	−131.11	−155.2	−104	− 75.56	+ 24.8	− 4	−20.00
	−402	−241.11		−302	185.56	−331.6	−202	−130.00	−151.6	−102	− 74.44	+ 28.4	− 2	−18.89
												+ 32.0	0	−17.78

Interpolation																		
To °F	0.2	0.4	0.5	0.7	0.9	1.1	1.3	1.4	1.6	1.8	2.0	2.2	2.3	2.5	2.7	2.9	3.1	3.2	3.4
From	0.1	0.2	0.3	0.4	0.5	0.6	0.7	0.8	0.9	1.0	1.1	1.2	1.3	1.4	1.5	1.6	1.7	1.8	1.9
To °C	0.06	0.11	0.17	0.22	0.28	0.33	0.39	0.44	0.50	0.56	0.61	0.67	0.72	0.78	0.83	0.89	0.94	1.00	1.06

TEMPERATURE CONVERSION TABLES (Continued)

0° to +250°

TO °F	FROM	TO °C	TO °F	FROM	TO °C	TO °F	FROM	TO °C	TO °F	FROM	TO °C	TO °F	FROM	TO °C
+ 32.0	+ 0	−17.78	122.0	50	10.00	212.0	100	37.78	302.0	150	65.56	392.0	200	93.33
33.8	1	−17.22	123.8	51	10.56	213.8	101	38.33	303.8	151	66.11	393.8	201	93.89
35.6	2	−16.67	125.6	52	11.11	215.6	102	38.89	305.6	152	66.67	395.6	202	94.44
37.4	3	−16.11	127.4	53	11.67	217.4	103	39.44	307.4	153	67.22	397.4	203	95.00
39.2	4	−15.56	129.2	54	12.22	219.2	104	40.00	309.2	154	67.78	399.2	204	95.56
41.0	5	−15.00	131.0	55	12.78	221.0	105	40.56	311.0	155	68.33	401.0	205	96.11
42.8	6	−14.44	132.8	56	13.33	222.8	106	41.11	312.8	156	68.89	402.8	206	96.67
44.6	7	−13.89	134.6	57	13.89	224.6	107	41.67	314.6	157	69.44	404.6	207	97.22
46.4	8	−13.33	136.4	58	14.44	226.4	108	42.22	316.4	158	70.00	406.4	208	97.78
48.2	9	−12.78	138.2	59	15.00	228.2	109	42.78	318.2	159	70.56	408.2	209	98.33
50.0	10	−12.22	140.0	60	15.56	230.0	110	43.33	320.0	160	71.11	410.0	210	98.89
51.8	11	−11.67	141.8	61	16.11	231.8	111	43.89	321.8	161	71.67	411.8	211	99.44
53.6	12	−11.11	143.6	62	16.67	233.6	112	44.44	323.6	162	72.22	413.6	212	100.00
55.4	13	−10.56	145.4	63	17.22	235.4	113	45.00	325.4	163	72.78	415.4	213	100.56
57.2	14	−10.00	147.2	64	17.78	237.2	114	45.56	327.2	164	73.33	417.2	214	101.11
59.0	15	− 9.44	149.0	65	18.33	239.0	115	46.11	329.0	165	73.89	419.0	215	101.67
60.8	16	− 8.89	150.8	66	18.89	240.8	116	46.67	330.8	166	74.44	420.8	216	102.22
62.6	17	− 8.33	152.6	67	19.44	242.6	117	47.22	332.6	167	75.00	422.6	217	102.78
64.4	18	− 7.78	154.4	68	20.00	244.4	118	47.78	334.4	168	75.56	424.4	218	103.33
66.2	19	− 7.22	156.2	69	20.56	246.2	119	48.33	336.2	169	76.11	426.2	219	103.89
68.0	20	− 6.67	158.0	70	21.11	248.0	120	48.89	338.0	170	76.67	428.0	220	104.44
69.8	21	− 6.11	159.8	71	21.67	249.8	121	49.44	339.8	171	77.22	429.8	221	105.00
71.6	22	− 5.56	161.6	72	22.22	251.6	122	50.00	341.6	172	77.78	431.6	222	105.56
73.4	23	− 5.00	163.4	73	22.78	253.4	123	50.56	343.4	173	78.33	433.4	223	106.11
75.2	24	− 4.44	165.2	74	23.33	255.2	124	51.11	345.2	174	78.89	435.2	224	106.67
77.0	25	− 3.89	167.0	75	23.89	257.0	125	51.67	347.0	175	79.44	437.0	225	107.22
78.0	26	− 3.33	168.8	76	24.44	258.8	126	52.22	348.8	176	80.00	438.8	226	107.78
80.6	27	− 2.78	170.6	77	25.00	260.6	127	52.78	350.6	177	80.56	440.6	227	108.33
82.4	28	− 2.22	172.4	78	25.56	262.4	128	53.33	352.4	178	81.11	442.4	228	108.89
84.2	29	− 1.67	174.2	79	26.11	264.2	129	53.89	354.2	179	81.67	444.2	229	109.44
86.0	30	− 1.11	176.0	80	26.67	266.0	130	54.44	356.0	180	82.22	446.0	230	110.00
87.8	31	− 0.56	177.8	81	27.22	267.8	131	55.00	357.8	181	82.78	447.8	231	110.56
89.6	32	0.00	179.6	82	27.78	269.6	132	55.56	359.8	182	83.33	449.6	232	111.11
91.4	33	+ 0.56	181.4	83	28.33	271.4	133	56.11	361.4	183	83.89	451.4	233	111.67
93.2	34	+ 1.11	183.2	84	28.89	273.2	134	56.67	363.2	184	84.44	453.2	234	112.22
95.0	35	1.67	185.0	85	29.44	275.0	135	57.22	365.0	185	85.00	455.0	235	112.78
96.8	36	2.22	186.8	86	30.00	276.8	136	57.78	366.8	186	85.56	456.8	236	113.33
98.6	37	2.78	188.6	87	30.56	278.6	137	58.33	368.6	187	86.11	458.6	237	113.89
100.4	38	3.33	190.4	88	31.11	280.4	138	58.89	370.4	188	86.67	460.4	238	114.44
102.2	39	3.89	192.2	89	31.67	282.2	139	59.44	372.2	189	87.22	462.2	239	115.00
104.4	40	4.44	194.0	90	32.22	284.0	140	60.00	374.0	190	87.78	464.0	240	115.56
105.8	41	5.00	195.8	91	32.78	285.8	141	60.56	375.8	191	88.33	465.8	241	116.11
107.6	42	5.56	197.6	92	33.33	287.6	142	61.11	377.6	192	88.89	467.6	242	116.67
109.4	43	6.11	199.4	93	33.89	289.4	143	61.67	379.4	193	89.44	469.4	243	117.22
111.2	44	6.67	201.2	94	34.44	291.2	144	62.22	381.2	194	90.00	471.2	244	117.78
113.0	45	7.22	203.0	95	35.00	293.0	145	62.78	383.0	195	90.56	473.0	245	118.33
114.8	46	7.78	204.8	96	35.56	294.8	146	63.33	384.8	196	91.11	474.8	246	118.89
116.6	47	8.33	206.6	97	36.11	296.6	147	63.89	386.6	197	91.67	476.6	247	119.44
118.4	48	8.89	208.4	98	36.67	298.4	148	64.44	388.4	198	92.22	478.4	248	120.00
120.2	49	9.44	210.2	99	37.22	300.2	149	65.00	390.2	199	92.78	480.2	249	120.56
												482.0	250	121.11

Interpolation

To °F	0.2	0.4	0.5	0.7	0.9	1.1	1.3	1.4	1.6
From	0.1	0.2	0.3	0.4	0.5	0.6	0.7	0.8	0.9
To °C	0.06	0.11	0.17	0.22	0.28	0.33	0.39	0.44	0.50

TEMPERATURE CONVERSION TABLES (Continued)

+250° to +500°

TO °F	FROM	TO °C	TO °F	FROM	TO °C	TO °F	FROM	TO °C	TO °F	FROM	TO °C	TO °F	FROM	TO °C
482.0	250	121.11	572.0	300	148.89	662.0	350	176.67	752.0	400	204.44	842.0	450	232.22
483.8	251	121.67	573.8	301	149.44	663.8	351	177.22	753.8	401	205.00	843.8	451	232.78
485.6	252	122.22	575.6	302	150.00	665.6	352	177.78	755.6	402	205.56	845.6	452	233.33
487.4	253	122.78	577.4	303	150.56	667.4	353	178.33	757.4	403	206.11	847.4	453	233.89
489.2	254	123.33	579.2	304	151.11	669.2	354	178.89	759.2	404	206.67	849.2	454	234.44
491.0	255	123.89	581.0	305	151.67	671.0	355	179.44	761.0	405	207.22	851.0	455	235.00
492.8	256	124.44	582.8	306	152.22	672.8	356	180.00	762.8	406	207.78	852.8	456	235.56
494.6	257	125.00	584.6	307	152.78	674.6	357	180.56	764.6	407	208.33	854.6	457	236.11
496.4	258	125.56	586.4	308	153.33	676.4	358	181.11	766.4	408	208.89	856.4	458	236.67
498.2	259	126.11	588.2	309	153.89	678.2	359	181.67	768.2	409	209.44	858.2	459	237.22
500.0	260	126.67	590.0	310	154.44	680.0	360	182.22	770.0	410	210.00	860.0	460	237.78
501.8	261	127.22	591.8	311	155.00	681.8	361	182.78	771.8	411	210.56	861.8	461	238.33
503.6	262	127.78	593.6	312	155.56	683.6	362	183.33	773.6	412	211.11	863.6	462	238.89
505.4	263	128.33	595.4	313	156.11	685.4	363	183.89	775.4	413	211.67	865.4	463	239.44
507.2	264	128.89	597.2	314	156.67	687.2	364	184.44	777.2	414	212.22	867.2	464	240.00
509.0	265	129.44	599.0	315	157.22	689.0	365	185.00	779.0	415	212.78	869.0	465	240.56
510.8	266	130.00	600.8	316	157.78	690.8	366	185.56	780.8	416	213.33	870.8	466	241.11
512.6	267	130.56	602.6	317	158.33	692.6	367	186.11	782.6	417	213.89	872.6	467	241.67
514.4	268	131.11	604.4	318	158.89	694.4	368	186.67	784.4	418	214.44	874.4	468	242.22
516.2	269	131.67	606.2	319	159.44	696.2	369	187.22	786.2	419	215.00	876.2	469	242.78
518.0	270	132.22	608.0	320	160.00	698.0	370	187.78	788.0	420	215.56	878.0	470	243.33
519.8	271	132.78	609.8	321	160.56	699.8	371	188.33	789.8	421	216.11	879.8	471	243.89
521.6	272	133.33	611.6	322	161.11	701.6	372	188.89	791.6	422	216.67	881.6	472	244.44
523.4	273	133.89	613.4	323	161.67	703.4	373	189.44	793.4	423	217.22	883.4	473	245.00
525.2	274	134.44	615.2	324	162.22	705.2	374	190.00	795.2	424	217.78	885.2	474	245.56
527.0	275	135.00	617.0	325	162.78	707.0	375	190.56	797.0	425	218.33	887.0	475	246.11
528.8	276	135.56	618.8	326	163.33	708.8	376	191.11	798.8	426	218.89	888.8	476	246.67
530.6	277	136.11	620.6	327	163.89	710.6	377	191.67	800.6	427	219.44	890.6	477	247.22
532.4	278	136.67	622.4	328	164.44	712.4	378	192.22	802.4	428	220.00	892.4	478	247.78
534.2	279	137.22	624.2	329	165.00	714.2	379	192.78	804.2	429	220.56	894.2	479	248.33
536.0	280	137.78	626.0	330	165.56	716.0	380	193.33	806.0	430	221.11	896.0	480	248.89
537.8	281	138.33	627.8	331	166.11	717.8	381	193.89	807.8	431	221.67	897.8	481	249.44
539.6	282	138.89	629.6	332	166.67	719.6	382	194.44	809.6	432	222.22	899.6	482	250.00
541.4	283	139.44	631.4	333	167.22	721.4	383	195.00	811.4	433	222.78	901.4	483	250.56
543.2	284	140.00	633.2	334	167.78	723.2	384	195.56	813.2	434	223.33	903.2	484	251.11
545.0	285	140.56	635.0	335	168.33	725.0	385	196.11	815.0	435	223.89	905.0	485	251.67
546.8	286	141.11	636.8	336	168.89	726.8	386	196.67	816.8	436	224.44	906.8	486	252.22
548.6	287	141.67	638.6	337	169.44	728.6	387	197.22	818.6	437	225.00	908.6	487	252.78
550.4	288	142.22	640.4	338	170.00	730.4	388	197.78	820.4	438	225.56	910.4	488	253.33
552.2	289	142.78	642.2	339	170.56	732.2	389	198.33	822.2	439	226.11	912.2	489	253.89
554.0	290	143.33	644.0	340	171.11	734.0	390	198.89	824.0	440	226.67	914.0	490	254.44
555.8	291	143.89	645.8	341	171.67	735.8	391	199.44	825.8	441	227.22	915.8	491	255.00
557.6	292	144.44	647.6	342	172.22	737.6	392	200.00	827.6	442	227.78	917.6	492	255.56
559.4	293	145.00	649.4	343	172.78	739.4	393	200.56	829.4	443	228.33	919.4	493	256.11
561.2	294	145.56	651.2	344	176.33	741.2	394	201.11	831.2	444	228.89	921.2	494	256.67
563.0	295	146.11	653.0	345	173.89	743.0	395	201.67	833.0	445	229.44	923.0	495	257.22
564.8	296	146.67	654.8	346	174.41	744.8	396	202.22	834.8	446	230.00	924.8	496	257.78
566.6	297	147.22	656.6	347	175.00	746.6	397	202.78	836.6	447	230.56	926.6	497	258.33
568.4	298	147.78	658.4	348	175.56	748.4	398	203.33	838.4	448	231.11	928.4	498	258.89
570.2	299	148.33	660.2	349	176.11	750.2	399	203.89	840.2	449	231.67	930.2	499	259.44
												932.0	500	260.00

Interpolation									
To °F	0.2	0.4	0.5	0.7	0.9	1.1	1.3	1.4	1.6
From	0.1	0.2	0.3	0.4	0.5	0.6	0.7	0.8	0.9
To °C	0.06	0.11	0.17	0.22	0.28	0.33	0.39	0.44	0.50

+ 500° to + 1750° **TEMPERATURE CONVERSION TABLES** (Continued)

TO °F	FROM	TO °C	TO °F	FROM	TO °C	TO °F	FROM	TO °C	TO °F	FROM	TO °C	TO °F	FROM	TO °C
932.0	500	260.00	1382.0	750	398.89	1832.0	1000	537.78	2282.0	1250	676.67	2732.0	1500	815.56
941.0	505	262.78	1391.0	755	401.67	1841.0	1005	540.56	2291.0	1255	679.44	2741.0	1505	818.33
950.0	510	265.56	1400.0	750	404.44	1850.0	1010	543.33	2300.0	1260	682.22	2750.0	1510	821.11
959.0	515	268.33	1409.0	765	407.22	1859.0	1015	546.11	2309.0	1265	685.00	2759.0	1515	823.89
968.0	520	271.11	1418.0	770	410.00	1868.0	1020	548.89	2318.0	1270	687.78	2768.0	1520	826.67
977.0	525	273.89	1427.0	775	412.78	1877.0	1025	551.67	2327.0	1275	690.56	2777.0	1525	829.44
986.0	530	276.67	1436.0	780	415.56	1886.0	1030	554.44	2336.0	1280	693.33	2786.0	1530	832.22
995.0	535	279.44	1445.0	785	418.33	1895.0	1035	557.22	2345.0	1285	696.11	2795.0	1535	835.00
1004.0	540	282.22	1454.0	790	421.11	1904.0	1040	560.00	2354.0	1290	698.89	2804.0	1540	837.78
1013.0	545	285.00	1463.0	795	423.89	1913.0	1045	562.78	2363.0	1295	701.67	2813.0	1545	840.56
1022.0	550	287.78	1472.0	800	426.67	1922.0	1050	565.56	2372.0	1300	704.44	2822.0	1550	843.33
1031.0	555	290.56	1481.0	805	429.44	1931.0	1055	568.33	2381.0	1305	707.22	2831.0	1555	846.11
1040.0	560	293.33	1490.0	810	432.22	1940.0	1060	571.11	2390.0	1310	710.00	2840.0	1560	848.89
1049.0	565	296.11	1499.0	815	435.00	1949.0	1065	573.89	2399.0	1315	712.78	2849.0	1565	851.67
1058.0	570	298.89	1508.0	820	437.78	1958.0	1070	576.67	2408.0	1320	715.56	2858.0	1570	854.44
1067.0	575	301.67	1517.0	825	440.56	1967.0	1075	579.44	2417.0	1325	178.33	2867.0	1575	857.22
1076.0	580	304.44	1526.0	830	443.33	1976.0	1080	582.22	2426.0	1330	721.11	2876.0	1580	830.00
1085.0	585	307.22	1535.0	835	446.11	1985.0	1085	585.00	2435.0	1335	723.89	2885.0	1585	862.78
1094.0	59ᴏ	310.00	1544.0	840	448.89	1994.0	1090	587.78	2444.0	1340	726.67	2894.0	1590	8Cᴏ56
1103.0	595	312.78	1553.0	845	451.67	2003.0	1095	590.55	2453.0	1345	729.44	2903.0	1595	868.33
1112.0	600	315.56	1562.0	850	454.44	2012.0	1100	593.33	2462.0	1350	732.22	2912.0	1600	871.11
1121.0	605	318.33	1571.0	855	457.22	2021.0	1105	596.11	2471.0	1355	735.00	2921.0	1605	873.89
1130.0	610	321.11	1580.0	860	460.00	2030.0	1110	598.89	2480.0	1360	737.78	2930.0	1610	876.67
1139.0	615	323.89	1589.0	865	462.78	2039.0	1115	601.67	2489.0	1365	740.56	2939.0	1615	879.44
1148.0	620	326.67	1598.0	870	465.56	2048.0	1120	604.44	2498.0	1370	743.33	2948.0	1620	882.22
1157.0	625	329.44	1607.0	875	468.33	2057.0	1125	607.22	2507.0	1375	746.11	2957.0	1625	885.00
1166.0	630	332.22	1616.0	880	471.11	2066.0	1130	610.00	2516.0	1380	748.89	2966.0	1630	887.78
1175.0	635	335.00	1625.0	885	473.89	2075.0	1135	612.78	2525.0	1385	751.67	2975.0	1635	890.56
1184.0	640	337.78	1634.0	890	476.67	2084.0	1140	615.56	2534.0	1390	754.44	2984.0	1640	893.33
1193.0	645	340.56	1643.0	895	479.44	2093.0	1145	618.33	2543.0	1395	757.22	2993.0	1645	896.11
1202.0	650	343.33	1652.0	900	482.22	2102.0	1150	621.11	2552.0	1400	760.00	3002.0	1650	898.89
1211.0	655	346.11	1661.0	905	485.00	2110.0	1155	623.89	2561.0	1405	762.78	3011.0	1655	901.67
1220.0	660	348.89	1670.0	910	487.78	2120.0	1160	626.67	2570.0	1410	765.56	3020.0	1660	904.44
1229.0	665	351.67	1679.0	915	490.56	2129.0	1165	629.44	2579.0	1415	768.33	3029.0	1665	907.22
1238.0	670	354.44	1668.0	920	493.33	2138.0	1170	632.22	2588.0	1420	771.11	3038.0	1670	910.00
1247.0	675	357.22	1697.0	925	496.11	2147.0	1175	635.00	2592.0	1425	773.89	3047.0	1675	912.78
1256.0	680	360.00	1706.0	930	498.89	2156.0	1180	637.78	2606.0	1430	776.67	3056.0	1680	915.56
1265.0	685	362.78	1715.0	935	501.67	2165.0	1185	640.55	2615.0	1435	779.44	3065.0	1685	918.33
1274.0	690	365.56	1724.0	940	504.44	2174.0	1190	643.33	2624.0	1440	782.22	3074.0	1690	921.11
1283.0	695	368.33	1733.0	945	507.22	2183.0	1195	646.11	2633.0	1445	785.00	3083.0	1695	923.89
1292.0	700	371.11	1742.0	950	510.00	2192.0	1200	648.89	2642.0	1450	787.78	3092.0	1700	926.67
1301.0	705	373.89	1751.0	955	512.78	2201.0	1205	651.67	2651.0	1455	790.56	3101.0	1705	929.44
1310.0	710	376.67	1760.0	960	515.56	2210.0	1210	654.44	2660.0	1460	793.33	3110.0	1710	932.22
1319.0	715	379.44	1769.0	965	518.33	2219.0	1215	657.22	2669.0	1465	796.11	3119.0	1715	935.00
1328.0	720	382.22	1778.0	970	521.11	2228.0	1220	660.00	2678.0	1470	798.89	3128.0	1720	937.78
1337.0	725	385.00	1787.0	975	523.89	2237.0	1225	662.78	2687.0	1475	801.67	3137.0	1725	940.56
1346.0	730	387.78	1796.0	980	526.67	2246.0	1230	665.56	2696.0	1480	804.44	3146.0	1730	943.33
1355.0	735	390.56	1805.0	985	529.44	2255.0	1235	668.33	2705.0	1485	807.22	3155.0	1735	946.11
1364.0	740	393.33	1814.0	990	532.22	2264.0	1240	671.11	2714.0	1490	810.00	3164.0	1740	948.89
1373.0	745	396.11	1823.0	995	535.00	2273.0	1245	673.89	2723.0	1495	812.78	3873.0	1745	951.67
												3182.0	1750	954.44

Interpolation

To °F	0.9	1.8	2.7	3.6	4.5	5.4	6.3	7.2	8.1
From	0.5	1.0	1.5	2.0	2.5	3.0	3.5	4.0	4.5
To °C	0.28	0.56	0.83	1.11	1.39	1.67	1.94	2.22	2.50

°API GRAVITY, SPECIFIC GRAVITY (60°/60°)
AND POUNDS PER GALLON

Hydrometer readings in degrees API are related by definition to specific gravity by the following formulas:

$$°API = \frac{141.5}{Sp.\ Gr.} - 131.5$$

$$Sp.\ Gr. = \frac{141.5}{°API + 131.5}$$

These relationships are for hydrometer readings made at or corrected to 60°F. and for specific gravity when defined as the ratio of the weight of a given volume of oil at 60°F. to the weight of the same volume of water at 60°F., all weighings being reduced to weights in vacuum by correcting for the buoyancy of air.

Pounds per gallon values are given as weight in air since commercial weighings are required by United States law to be on this basis. For engineering design purposes, the difference between weight in air and weight in vacuum will generally be trivial.

Tables have been published[1,2] covering the conversion of these units over the range of 0° to 100° API. For convenience in labeling petroleum materials and correlating their properties it is convenient to have conversions over a wider range. The conversions were recalculated by a digital computer program over the range of −20° to 300° API using the procedure for buoyancy corrections outlined by the Bureau of Standards. The recalculated values checked the Bureau of Standards and the ASTM-IP tables over the published range with occasional discrepancies of one unit in the last decimal place.

The Bureau of Standards and the ASTM-IP tables include conversion tables for the effect of temperature on API gravity and volumes.

1. *National Standard Petroleum Oil Tables. Circular of the National Bureau of Standards C410. (1936) Supplement to NBS Circular C410* (April 20, 1937).

2. *ASTM-IP Petroleum Measurement Tables (American Edition)* American Society for Testing Materials (1952).

°API GRAVITY, SPECIFIC GRAVITY (60°/60°) AND POUNDS PER GALLON

−20 TO 5°API

°API	SPECIFIC GRAVITY	POUNDS PER GALLON	°API	SPECIFIC GRAVITY	POUNDS PER GALLON	°API	SPECIFIC GRAVITY	POUNDS PER GALLON	°API	SPECIFIC GRAVITY	POUNDS PER GALLON	°API	SPECIFIC GRAVITY	POUNDS PER GALLON
−20.0	1.2691	10.572	−15.0	1.2146	10.118	−10.0	1.1646	9.701	−5.0	1.1186	9.317	0.0	1.0760	8.962
−19.9	1.2679	10.562	−14.9	1.2136	10.109	−9.9	1.1637	9.693	−4.9	1.1177	9.310	0.1	1.0752	8.956
−19.8	1.2668	10.553	−14.8	1.2125	10.100	−9.8	1.1627	9.685	−4.8	1.1168	9.302	0.2	1.0744	8.949
−19.7	1.2657	10.543	−14.7	1.2115	10.092	−9.7	1.1617	9.677	−4.7	1.1159	9.295	0.3	1.0736	8.942
−19.6	1.2645	10.534	−14.6	1.2104	10.083	−9.6	1.1608	9.669	−4.6	1.1151	9.288	0.4	1.0728	8.935
−19.5	1.2634	10.524	−14.5	1.2094	10.074	−9.5	1.1598	9.661	−4.5	1.1142	9.280	0.5	1.0720	8.928
−19.4	1.2623	10.515	−14.4	1.2084	10.066	−9.4	1.1589	9.653	−4.4	1.1133	9.273	0.6	1.0712	8.922
−19.3	1.2611	10.506	−14.3	1.2073	10.057	−9.3	1.1579	9.645	−4.3	1.1124	9.266	0.7	1.0703	8.915
−19.2	1.2600	10.496	−14.2	1.2063	10.048	−9.2	1.1570	9.637	−4.2	1.1115	9.258	0.8	1.0695	8.908
−19.1	1.2589	10.487	−14.1	1.2053	10.040	−9.1	1.1560	9.629	−4.1	1.1107	9.251	0.9	1.0687	8.901
−19.0	1.2578	10.478	−14.0	1.2043	10.031	−9.0	1.1551	9.622	−4.0	1.1098	9.244	1.0	1.0679	8.895
−18.9	1.2567	10.468	−13.9	1.2032	10.023	−8.9	1.1542	9.614	−3.9	1.1089	9.237	1.1	1.0671	8.888
−18.8	1.2555	10.459	−13.8	1.2022	10.014	−8.8	1.1532	9.606	−3.8	1.1081	9.229	1.2	1.0663	8.881
−18.7	1.2544	10.450	−13.7	1.2012	10.006	−8.7	1.1523	9.598	−3.7	1.1072	9.222	1.3	1.0655	8.875
−18.6	1.2533	10.440	−13.6	1.2002	9.997	−8.6	1.1513	9.590	−3.6	1.1063	9.215	1.4	1.0647	8.868
−18.5	1.2522	10.431	−13.5	1.1992	9.989	−8.5	1.1504	9.582	−3.5	1.1055	9.208	1.5	1.0639	8.861
−18.4	1.2511	10.422	−13.4	1.1981	9.980	−8.4	1.1495	9.575	−3.4	1.1046	9.200	1.6	1.0631	8.855
−18.3	1.2500	10.413	−13.3	1.1971	9.972	−8.3	1.1485	9.567	−3.3	1.1037	9.193	1.7	1.0623	8.848
−18.2	1.2489	10.404	−13.2	1.1961	9.963	−8.2	1.1476	9.559	−3.2	1.1029	9.186	1.8	1.0615	8.841
−18.1	1.2478	10.394	−13.1	1.1951	9.955	−8.1	1.1467	9.551	−3.1	1.1020	9.179	1.9	1.0607	8.835
−18.0	1.2467	10.385	−13.0	1.1941	9.947	−8.0	1.1457	9.544	−3.0	1.1012	9.172	2.0	1.0599	8.828
−17.9	1.2456	10.376	−12.9	1.1931	9.938	−7.9	1.1448	9.536	−2.9	1.1003	9.165	2.1	1.0591	8.821
−17.8	1.2445	10.367	−12.8	1.1921	9.930	−7.8	1.1439	9.528	−2.8	1.0995	9.158	2.2	1.0583	8.815
−17.7	1.2434	10.358	−12.7	1.1911	9.921	−7.7	1.1430	9.520	−2.7	1.0986	9.150	2.3	1.0575	8.808
−17.6	1.2423	10.349	−12.6	1.1901	9.913	−7.6	1.1421	9.513	−2.6	1.0978	9.143	2.4	1.0568	8.802
−17.5	1.2412	10.340	−12.5	1.1891	9.905	−7.5	1.1411	9.505	−2.5	1.0969	9.136	2.5	1.0560	8.795
−17.4	1.2401	10.331	−12.4	1.1881	9.896	−7.4	1.1402	9.497	−2.4	1.0961	9.129	2.6	1.0552	8.788
−17.3	1.2391	10.322	−12.3	1.1871	9.888	−7.3	1.1393	9.490	−2.3	1.0952	9.122	2.7	1.0544	8.782
−17.2	1.2380	10.312	−12.2	1.1861	9.880	−7.2	1.1384	9.482	−2.2	1.0944	9.115	2.8	1.0536	8.775
−17.1	1.2369	10.303	−12.1	1.1851	9.872	−7.1	1.1375	9.474	−2.1	1.0935	9.108	2.9	1.0528	8.769
−17.0	1.2358	10.294	−12.0	1.1841	9.863	−7.0	1.1365	9.467	−2.0	1.0927	9.101	3.0	1.0520	8.762
−16.9	1.2347	10.285	−11.9	1.1831	9.855	−6.9	1.1356	9.459	−1.9	1.0918	9.094	3.1	1.0513	8.756
−16.8	1.2337	10.276	−11.8	1.1821	9.847	−6.8	1.1347	9.452	−1.8	1.0910	9.087	3.2	1.0505	8.749
−16.7	1.2326	10.268	−11.7	1.1811	9.839	−6.7	1.1338	9.444	−1.7	1.0901	9.080	3.3	1.0497	8.743
−16.6	1.2315	10.259	−11.6	1.1802	9.830	−6.6	1.1329	9.437	−1.6	1.0893	9.073	3.4	1.0489	8.736
−16.5	1.2304	10.250	−11.5	1.1792	9.822	−6.5	1.1320	9.429	−1.5	1.0885	9.066	3.5	1.0481	8.730
−16.4	1.2294	10.241	−11.4	1.1782	9.814	−6.4	1.1311	9.421	−1.4	1.0876	9.059	3.6	1.0474	8.723
−16.3	1.2283	10.232	−11.3	1.1772	9.806	−6.3	1.1302	9.414	−1.3	1.0868	9.052	3.7	1.0466	8.717
−16.2	1.2272	10.223	−11.2	1.1762	9.798	−6.2	1.1293	9.406	−1.2	1.0860	9.045	3.8	1.0458	8.710
−16.1	1.2262	10.214	−11.1	1.1752	9.790	−6.1	1.1284	9.399	−1.1	1.0851	9.038	3.9	1.0451	8.704
−16.0	1.2251	10.205	−11.0	1.1743	9.781	−6.0	1.1275	9.391	−1.0	1.0843	9.031	4.0	1.0443	8.697
−15.9	1.2240	10.196	−10.9	1.1733	9.773	−5.9	1.1266	9.384	−0.9	1.0835	9.024	4.1	1.0435	8.691
−15.8	1.2230	10.188	−10.8	1.1723	9.765	−5.8	1.1257	9.376	−0.8	1.0826	9.017	4.2	1.0427	8.685
−15.7	1.2219	10.179	−10.7	1.1714	9.757	−5.7	1.1248	9.369	−0.7	1.0818	9.010	4.3	1.0420	8.678
−15.6	1.2209	10.170	−10.6	1.1704	9.749	−5.6	1.1239	9.362	−0.6	1.0810	9.003	4.4	1.0412	8.672
−15.5	1.2198	10.161	−10.5	1.1694	9.741	−5.5	1.1230	9.354	−0.5	1.0802	8.997	4.5	1.0404	8.665
−15.4	1.2188	10.152	−10.4	1.1685	9.733	−5.4	1.1221	9.347	−0.4	1.0793	8.990	4.6	1.0397	8.659
−15.3	1.2177	10.144	−10.3	1.1675	9.725	−5.3	1.1212	9.339	−0.3	1.0785	8.983	4.7	1.0389	8.653
−15.2	1.2167	10.135	−10.2	1.1665	9.717	−5.2	1.1203	9.332	−0.2	1.0777	8.976	4.8	1.0382	9.646
−15.1	1.2156	10.126	−10.1	1.1656	9.709	−5.1	1.1195	9.324	−0.1	1.0769	8.969	4.9	1.0374	8.640
												5.0	1.0366	8.634

°API GRAVITY, SPECIFIC GRAVITY (60°/60°) AND POUNDS PER GALLON (Continued)

5 TO 30°API

°API	SPECIFIC GRAVITY	POUNDS PER GALLON	°API	SPECIFIC GRAVITY	POUNDS PER GALLON	°API	SPECIFIC GRAVITY	POUNDS PER GALLON	°API	SPECIFIC GRAVITY	POUNDS PER GALLON	°API	SPECIFIC GRAVITY	POUNDS PER GALLON	°API	SPECIFIC GRAVITY	POUNDS PER GALLON
5.0	1.0366	8.634	10.0	1.0000	8.328	15.0	0.9659	8.044	20.0	0.9340	7.778	25.0	0.9042	7.529
5.1	1.0359	8.627	10.1	0.9993	8.322	15.1	0.9652	8.038	20.1	0.9334	7.773	25.1	0.9036	7.524
5.2	1.0351	8.621	10.2	0.9986	8.317	15.2	0.9646	8.033	20.2	0.9328	7.768	25.2	0.9030	7.519
5.3	1.0344	8.615	10.3	0.9979	8.311	15.3	0.9639	8.027	20.3	0.9321	7.762	25.3	0.9024	7.515
5.4	1.0336	8.608	10.4	0.9972	8.305	15.4	0.9632	8.022	20.4	0.9315	7.757	25.4	0.9018	7.510
5.5	1.0328	8.602	10.5	0.9965	8.299	15.5	0.9626	8.016	20.5	0.9309	7.752	25.5	0.9013	7.505
5.6	1.0321	8.596	10.6	0.9958	8.293	15.6	0.9619	8.011	20.6	0.9303	7.747	25.6	0.9007	7.500
5.7	1.0313	8.590	10.7	0.9951	8.287	15.7	0.9613	8.005	20.7	0.9297	7.742	25.7	0.9001	7.495
5.8	1.0306	8.583	10.8	0.9944	8.281	15.8	0.9606	8.000	20.8	0.9291	7.737	25.8	0.8996	7.491
5.9	1.0298	8.577	10.9	0.9937	8.276	15.9	0.9600	7.995	20.9	0.9285	7.732	25.9	0.8990	7.486
6.0	1.0291	8.571	11.0	0.9930	8.270	16.0	0.9593	7.989	21.0	0.9279	7.727	26.0	0.8984	7.481
6.1	1.0283	8.565	11.1	0.9923	8.264	16.1	0.9587	7.984	21.1	0.9273	7.722	26.1	0.8978	7.476
6.2	1.0276	8.558	11.2	0.9916	8.258	16.2	0.9580	7.978	21.2	0.9267	7.717	26.2	0.8973	7.472
6.3	1.0269	8.552	11.3	0.9909	8.252	16.3	0.9574	7.973	21.3	0.9260	7.712	26.3	0.8967	7.467
6.4	1.0261	8.546	11.4	0.9902	8.247	16.4	0.9567	7.967	21.4	0.9254	7.707	26.4	0.8961	7.462
6.5	1.0254	8.540	11.5	0.9895	8.241	16.5	0.9561	7.962	21.5	0.9248	7.702	26.5	0.8956	7.457
6.6	1.0246	8.534	11.6	0.9888	8.235	16.6	0.9554	7.957	21.6	0.9242	7.696	26.6	0.8950	7.453
6.7	1.0239	8.257	11.7	0.9881	8.229	16.7	0.9548	7.951	21.7	0.9236	7.691	26.7	0.8944	7.448
6.8	1.0231	8.521	11.8	0.9874	8.221	16.8	0.9541	7.946	21.8	0.9230	7.686	26.8	0.8939	7.443
6.9	1.0224	8.515	11.9	0.9868	8.218	16.9	0.9535	7.941	21.9	0.9224	7.681	26.9	0.8933	7.439
7.0	1.0217	8.509	12.0	0.9861	8.212	17.0	0.9529	7.935	22.0	0.9218	7.676	27.0	0.8927	7.434
7.1	1.0209	8.503	12.1	0.9854	8.206	17.1	0.9522	7.930	22.1	0.9212	7.671	27.1	0.8922	7.429
7.2	1.0202	8.497	12.2	0.9847	8.201	17.2	0.9516	7.925	22.2	0.9206	7.666	27.2	0.8916	7.425
7.3	1.0195	8.490	12.3	0.9840	8.195	17.3	0.9509	7.919	22.3	0.9200	7.661	27.3	0.8911	7.420
7.4	1.0187	8.484	12.4	0.9833	8.189	17.4	0.9503	7.914	22.4	0.9194	7.656	27.4	0.8905	7.415
7.5	1.0180	8.478	12.5	0.9826	8.184	17.5	0.9497	7.909	22.5	0.9188	7.651	27.5	0.8899	7.411
7.6	1.0173	8.472	12.6	0.9820	8.178	17.6	0.9490	7.903	22.6	0.9182	7.646	27.6	0.8894	7.406
7.7	1.0165	8.466	12.7	0.9813	8.172	17.7	0.9484	7.898	22.7	0.9176	7.642	27.7	0.8888	7.401
7.8	1.0158	8.460	12.8	0.9806	8.166	17.8	0.9478	7.893	22.8	0.9170	7.637	27.8	0.8883	7.397
7.9	1.0151	8.454	12.9	0.9799	8.161	17.9	0.9471	7.887	22.9	0.9165	7.632	27.9	0.8877	7.392
8.0	1.0143	8.448	13.0	0.9792	8.155	18.0	0.9465	7.882	23.0	0.9159	7.627	28.0	0.8871	7.387
8.1	1.0136	8.442	13.1	0.9786	8.150	18.1	0.9459	7.877	23.1	0.9153	7.622	28.1	0.8866	7.383
8.2	1.0129	8.436	13.2	0.9779	8.144	18.2	0.9452	7.872	23.2	0.9147	7.617	28.2	0.8860	7.378
8.3	1.0122	8.430	13.3	0.9772	8.138	18.3	0.9446	7.866	23.3	0.9141	7.612	28.3	0.8855	7.373
8.4	1.0114	8.424	13.4	0.9765	8.133	18.4	0.9440	7.861	23.4	0.9135	7.607	28.4	0.8849	7.369
8.5	1.0107	8.418	13.5	0.9759	8.127	18.5	0.9433	7.856	23.5	0.9129	7.602	28.5	0.8844	7.364
8.6	1.0100	8.412	13.6	0.9752	8.121	18.6	0.9427	7.851	23.6	0.9123	7.597	28.6	0.8838	7.360
8.7	1.0093	8.406	13.7	0.9745	8.116	18.7	0.9421	7.845	23.7	0.9117	7.592	28.7	0.8833	7.355
8.8	1.0086	8.400	13.8	0.9738	8.110	18.8	0.9415	7.840	23.8	0.9111	7.587	28.8	0.8827	7.350
8.9	1.0078	8.394	13.9	0.9732	8.105	18.9	0.9408	7.835	23.9	0.9106	7.582	28.9	0.8822	7.346
9.0	1.0071	8.388	14.0	0.9725	8.099	19.0	0.9402	7.830	24.0	0.9100	7.578	29.0	0.8816	7.341
9.1	1.0064	8.382	14.1	0.9718	8.093	19.1	0.9396	7.824	24.1	0.9094	7.573	29.1	0.8811	7.337
9.2	1.0057	8.376	14.2	0.9712	8.088	19.2	0.9390	7.819	24.2	0.9088	7.568	29.2	0.8805	7.332
9.3	1.0050	8.370	14.3	0.9705	8.082	19.3	0.9383	7.814	24.3	0.9082	7.563	29.3	0.8800	7.327
9.4	1.0043	8.364	14.4	0.9698	8.077	19.4	0.9377	7.809	24.4	0.9076	7.558	29.4	0.8794	7.323
9.5	1.0035	8.358	14.5	0.9692	8.071	19.5	0.9371	7.804	24.5	0.9071	7.553	29.5	0.8789	7.318
9.6	1.0028	8.352	14.6	0.9685	8.066	19.6	0.9365	7.798	24.6	0.9065	7.548	29.6	0.8783	7.314
9.7	1.0021	8.346	14.7	0.9679	8.060	19.7	0.9358	7.793	24.7	0.9059	7.544	29.7	0.8778	7.309
9.8	1.0014	8.340	14.8	0.9672	8.055	19.8	0.9352	7.788	24.8	0.9053	7.539	29.8	0.8772	7.305
9.9	1.0007	8.334	14.9	0.9665	8.049	19.9	0.9346	7.783	24.9	0.9047	7.534	29.9	0.8767	7.300
												30.0	0.8762	7.296

°API GRAVITY, SPECIFIC GRAVITY (60°/60°)
AND POUNDS PER GALLON (Continued)

30 TO 55°API

°API	SPECIFIC GRAVITY	POUNDS PER GALLON	°API	SPECIFIC GRAVITY	POUNDS PER GALLON	°API	SPECIFIC GRAVITY	POUNDS PER GALLON	°API	SPECIFIC GRAVITY	POUNDS PER GALLON	°API	SPECIFIC GRAVITY	POUNDS PER GALLON
30.0	0.8762	7.296	35.0	0.8499	7.076	40.0	0.8251	6.870	45.0	0.8017	6.675	50.0	0.7796	6.491
30.1	0.8756	7.291	35.1	0.8493	7.072	40.1	0.8246	6.866	45.1	0.8012	6.671	50.1	0.7792	6.487
30.2	0.8751	7.287	35.2	0.8488	7.068	40.2	0.8241	6.862	45.2	0.8008	6.667	50.2	0.7788	6.483
30.3	0.8745	7.282	35.3	0.8483	7.064	40.3	0.8236	6.858	45.3	0.8003	6.663	50.3	0.7783	6.480
30.4	0.8740	7.278	35.4	0.8478	7.059	40.4	0.8232	6.854	45.4	0.7999	6.660	50.4	0.7779	6.476
30.5	0.8735	7.273	35.5	0.8473	7.055	40.5	0.8227	6.850	45.5	0.7994	6.656	50.5	0.7775	6.473
30.6	0.8729	7.269	35.6	0.8468	7.051	40.6	0.8222	6.846	45.6	0.7990	6.652	50.6	0.7770	6.469
30.7	0.8724	7.264	35.7	0.8463	7.047	40.7	0.8217	6.842	45.7	0.7985	6.648	50.7	0.7766	6.466
30.8	0.8718	7.260	35.8	0.8458	7.042	40.8	0.8212	6.838	45.8	0.7981	6.645	50.8	0.7762	6.462
30.9	0.8713	7.255	35.9	0.8453	7.038	40.9	0.8208	6.834	45.9	0.7976	6.641	50.9	0.7758	6.459
31.0	0.8708	7.251	36.0	0.8448	7.034	41.0	0.8203	6.830	46.0	0.7972	6.637	51.0	0.7753	6.455
31.1	0.8702	7.246	36.1	0.8443	7.030	41.1	0.8198	6.826	46.1	0.7967	6.633	51.1	0.7749	6.451
31.2	0.8697	7.242	36.2	0.8438	7.026	41.2	0.8193	6.822	46.2	0.7963	6.630	51.2	0.7745	6.448
31.3	0.8692	7.237	36.3	0.8433	7.021	41.3	0.8189	6.818	46.3	0.7958	6.626	51.3	0.7741	6.444
31.4	0.8686	7.233	36.4	0.8428	7.017	41.4	0.8184	6.814	46.4	0.7954	6.622	51.4	0.7736	6.441
31.5	0.8681	7.228	36.5	0.8423	7.013	41.5	0.8179	6.810	46.5	0.7949	6.618	51.5	0.7732	6.437
31.6	0.8676	7.224	36.6	0.8418	7.009	41.6	0.8174	6.806	46.6	0.7945	6.615	51.6	0.7728	6.434
31.7	0.8670	7.220	36.7	0.8413	7.005	41.7	0.8170	6.802	46.7	0.7941	6.611	51.7	0.7724	6.430
31.8	0.8665	7.215	36.8	0.8408	7.000	41.8	0.8165	6.798	46.8	0.7936	6.607	51.8	0.7720	6.427
31.9	0.8660	7.211	36.9	0.8403	6.996	41.9	0.8160	6.794	46.9	0.7932	6.604	51.9	0.7715	6.423
32.0	0.8654	7.206	37.0	0.8398	6.992	42.0	0.8156	6.790	47.0	0.7927	6.600	52.0	0.7711	6.420
32.1	0.8649	7.202	37.1	0.8393	6.988	42.1	0.8151	6.786	47.1	0.7923	6.596	52.1	0.7707	6.416
32.2	0.8644	7.197	37.2	0.8388	6.984	42.2	0.8146	6.783	47.2	0.7918	6.592	52.2	0.7703	6.413
32.3	0.8639	7.193	37.3	0.8383	6.980	42.3	0.8142	6.779	47.3	0.7914	6.589	52.3	0.7699	6.409
32.4	0.8633	7.189	37.4	0.8378	6.976	42.4	0.8137	6.775	47.4	0.7909	6.585	52.4	0.7694	6.406
32.5	0.8628	7.184	37.5	0.8373	6.971	42.5	0.8132	6.771	47.5	0.7905	6.581	52.5	0.7690	6.402
32.6	0.8623	7.180	37.6	0.8368	6.967	42.6	0.8128	6.767	47.6	0.7901	6.578	52.6	0.7686	6.399
32.7	0.8618	7.176	37.7	0.8363	6.963	42.7	0.8123	6.763	47.7	0.7896	6.574	52.7	0.7682	6.395
32.8	0.8612	7.171	37.8	0.8358	6.959	42.8	0.8118	6.759	47.8	0.7892	6.570	52.8	0.7678	6.392
32.9	0.8607	7.167	37.9	0.8353	6.955	42.9	0.8114	6.755	47.9	0.7887	6.567	52.9	0.7674	6.388
33.0	0.8602	7.162	38.0	0.8348	6.951	43.0	0.8109	6.751	48.0	0.7883	6.563	53.0	0.7669	6.385
33.1	0.8597	7.158	38.1	0.8343	6.947	43.1	0.8104	6.748	48.1	0.7879	6.559	53.1	0.7665	6.381
33.2	0.8591	7.154	38.2	0.8338	6.943	43.2	0.8100	6.744	48.2	0.7874	6.556	53.2	0.7661	6.378
33.3	0.8586	7.149	38.3	0.8333	6.939	43.3	0.8095	6.740	48.3	0.7870	6.552	53.3	0.7657	6.375
33.4	0.8581	7.145	38.4	0.8328	6.934	43.4	0.8090	6.736	48.4	0.7865	6.548	53.4	0.7563	6.371
33.5	0.8576	7.141	38.5	0.8324	6.930	43.5	0.8086	6.732	48.5	0.7861	6.545	53.5	0.7649	6.368
33.6	0.8571	7.136	38.6	0.8319	6.926	43.6	0.8081	6.728	48.6	0.7857	6.541	53.6	0.7645	6.364
33.7	0.8565	7.132	38.7	0.8314	6.922	43.7	0.8076	6.724	48.7	0.7852	6.538	53.7	0.7640	6.361
33.8	0.8560	7.128	38.8	0.8309	6.918	43.8	0.8072	6.721	48.8	0.7848	6.534	53.8	0.7636	6.357
33.9	0.8555	7.123	38.9	0.8304	6.914	43.9	0.8067	6.717	48.9	0.7844	6.530	53.9	0.7632	6.354
34.0	0.8550	7.119	39.0	0.8299	6.910	44.0	0.8063	6.713	49.0	0.7839	6.527	54.0	0.7628	6.350
34.1	0.8545	7.115	39.1	0.8294	6.906	44.1	0.8058	6.709	49.1	0.7835	6.523	54.1	0.7624	6.347
34.2	0.8540	7.110	39.2	0.8289	6.902	44.2	0.8054	6.705	49.2	0.7831	6.519	54.2	0.7620	6.344
34.3	0.8534	7.106	39.3	0.8285	6.898	44.3	0.8049	6.701	49.3	0.7826	6.516	54.3	0.7616	6.340
34.4	0.8529	7.102	39.4	0.8280	6.894	44.4	0.8044	6.698	49.4	0.7822	6.512	54.4	0.7612	6.337
34.5	0.8524	7.098	39.5	0.8275	6.890	44.5	0.8040	6.694	49.5	0.7818	6.509	54.5	0.7608	6.333
34.6	0.8519	7.093	39.6	0.8270	6.886	44.6	0.8035	6.690	49.6	0.7813	6.505	54.6	0.7603	6.330
34.7	0.8514	7.089	39.7	0.8265	6.882	44.7	0.8031	6.686	49.7	0.7809	6.501	54.7	0.7599	6.327
34.8	0.8509	7.085	39.8	0.8260	6.878	44.8	0.8026	6.682	49.8	0.7805	6.498	54.8	0.7595	6.323
34.9	0.8504	7.081	39.9	0.8256	6.874	44.9	0.8022	6.679	49.9	0.7800	6.494	54.9	0.7591	6.320
												55.0	0.7587	6.316

°API GRAVITY, SPECIFIC GRAVITY (60°/60°) AND POUNDS PER GALLON (Continued)

55 TO 80° API

°API	SPECIFIC GRAVITY	POUNDS PER GALLON	°API	SPECIFIC GRAVITY	POUNDS PER GALLON	°API	SPECIFIC GRAVITY	POUNDS PER GALLON	°API	SPECIFIC GRAVITY	POUNDS PER GALLON	°API	SPECIFIC GRAVITY	POUNDS PER GALLON
55.0	0.7587	6.316	60.0	0.7389	6.151	65.0	0.7201	5.994	70.0	0.7022	5.845	75.0	0.6852	5.704
55.1	0.7583	6.313	60.1	0.7385	6.148	65.1	0.7197	5.991	70.1	0.7019	5.842	75.1	0.6849	5.701
55.2	0.7579	6.310	60.2	0.7381	6.145	65.2	0.7194	5.988	70.2	0.7015	5.840	75.2	0.6846	5.698
55.3	0.7575	6.306	60.3	0.7377	6.141	65.3	0.7190	5.985	70.3	0.7012	5.837	75.3	0.6842	5.695
55.4	0.7571	6.303	60.4	0.7374	6.138	65.4	0.7186	5.982	70.4	0.7008	5.834	75.4	0.6839	5.693
55.5	0.7567	6.299	60.5	0.7370	6.135	65.5	0.7183	5.979	70.5	0.7005	5.831	75.5	0.6836	5.690
55.6	0.7563	6.296	60.6	0.7366	6.132	65.6	0.7179	5.976	70.6	0.7001	5.828	75.6	0.6832	5.687
55.7	0.7559	.6.293	60.7	0.7362	6.129	65.7	0.7175	5.973	70.7	0.6998	5.825	75.7	0.6829	5.684
55.8	0.7555	6.289	60.8	0.7358	6.126	65.8	0.7172	5.970	70.8	0.6995	5.822	75.8	0.6826	5.682
55.9	0.7551	6.286	60.9	0.7354	6.122	65.9	0.7168	5.967	70.9	0.6991	5.819	75.9	0.6823	5.679
56.0	0.7547	6.283	61.0	0.7351	6.119	66.0	0.7165	5.964	71.0	0.6988	5.816	76.0	0.6819	5.676
56.1	0.7543	6.279	61.1	0.7347	6.116	66.1	0.7161	5.961	71.1	0.6984	5.814	76.1	0.6816	5.673
56.2	0.7539	6.276	61.2	0.7343	6.113	66.2	0.7157	5.958	71.2	0.6981	5.811	76.2	0.6813	5.671
56.3	0.7535	6.273	61.3	0.7339	6.110	66.3	0.7154	5.955	71.3	0.6977	5.808	76.3	0.6809	5.668
56.4	0.7531	6.269	61.4	0.7335	6.106	66.4	0.7150	5.952	71.4	0.6974	5.805	76.4	0.6806	5.665
56.5	0.7527	6.266	61.5	0.7332	6.103	66.5	0.7146	5.949	71.5	0.6970	5.802	76.5	0.6803	5.662
56.6	0.7523	6.263	61.6	0.7328	6.100	66.6	0.7143	5.946	71.6	0.6967	5.799	76.6	0.6800	5.660
56.7	0.7519	6.259	61.7	0.7324	6.097	66.7	0.7139	5.943	71.7	0.6964	5.796	76.7	0.6796	5.657
56.8	0.7515	6.256	61.8	0.7320	6.094	66.8	0.7136	5.940	71.8	0.6960	5.794	76.8	0.6793	5.654
56.9	0.7511	6.253	61.9	0.7316	6.091	66.9	0.7132	5.937	71.9	0.6957	5.791	76.9	0.6790	5.651
57.0	0.7507	6.249	62.0	0.7313	6.087	67.0	0.7128	5.934	72.0	0.6953	5.788	77.0	0.6787	5.649
57.1	0.7503	6.246	62.1	0.7309	6.084	67.1	0.7125	5.931	72.1	0.6950	5.785	77.1	0.6783	5.646
57.2	0.7499	6.243	62.2	0.7305	6.081	67.2	0.7121	5.928	72.2	0.6946	5.782	77.2	0.6780	5.643
57.3	0.7495	6.239	62.3	0.7301	6.078	67.3	0.7118	5.925	72.3	0.6943	5.779	77.3	0.6777	5.641
57.4	0.7491	6.236	62.4	0.7298	6.075	67.4	0.7114	5.922	72.4	0.6940	5.776	77.4	0.6774	5.638
57.5	0.7487	6.233	62.5	0.7294	6.072	67.5	0.7111	5.919	72.5	0.6936	5.774	77.5	0.6770	5.635
57.6	0.7483	6.229	62.6	0.7290	6.069	67.6	0.7107	5.916	72.6	0.6933	5.771	77.6	0.6767	5.633
57.7	0.7479	6.226	62.7	0.7286	6.065	67.7	0.7103	5.913	72.7	0.6929	5.768	77.7	0.6764	5.630
57.8	0.7475	6.223	62.8	0.7283	6.062	67.8	0.7100	5.910	72.8	0.6926	5.765	77.8	0.6761	5.627
57.9	0.7471	6.219	62.9	0.7279	6.059	67.9	0.7096	5.907	72.9	0.6923	5.762	77.9	0.6757	5.624
58.0	0.7467	6.216	63.0	0.7275	6.056	68.0	0.7093	5.904	73.0	0.6919	5.759	78.0	0.6754	5.622
58.1	0.7463	6.213	63.1	0.7271	6.053	68.1	0.7089	5.901	73.1	0.6916	5.757	78.1	0.6751	5.619
58.2	0.7459	6.210	63.2	0.7268	6.050	68.2	0.7086	5.898	73.2	0.6913	5.754	78.2	0.6748	5.616
58.3	0.7455	6.206	63.3	0.7264	6.047	68.3	0.7082	5.895	73.3	0.6909	5.751	78.3	0.6745	5.614
58.4	0.7451	6.203	63.4	0.7260	6.044	68.4	0.7079	5.892	73.4	0.6906	5.748	78.4	0.6741	5.611
58.5	0.7447	6.200	63.5	0.7256	6.041	68.5	0.7075	5.889	73.5	0.6902	5.745	78.5	0.6738	5.608
58.6	0.7443	6.197	63.6	0.7253	6.037	68.6	0.7071	5.886	73.6	0.6899	5.743	78.6	0.6735	5.606
58.7	0.7440	6.193	63.7	0.7249	6.034	68.7	0.7068	5.883	73.7	0.6896	5.740	78.7	0.6732	5.603
58.8	0.7436	6.190	63.8	0.7245	6.031	68.8	0.7064	5.880	73.8	0.6892	5.737	78.8	0.6728	5.600
58.9	0.7432	6.187	63.9	0.7242	6.028	68.9	0.7061	5.878	73.9	0.6889	5.734	78.9	0.6725	5.598
59.0	0.7428	6.183	64.0	0.7238	6.025	69.0	0.7057	5.875	74.0	0.6886	5.731	79.0	0.6722	5.595
59.1	0.7424	6.180	64.1	0.7234	6.022	69.1	0.7054	5.872	74.1	0.6882	5.729	79.1	0.6719	5.592
59.2	0.7420	6.177	64.2	0.7230	6.019	69.2	0.7050	5.869	74.2	0.6879	5.726	79.2	0.6716	5.590
59.3	0.7416	6.174	64.3	0.7227	6.016	69.3	0.7047	5.866	74.3	0.6876	5.723	79.3	0.6713	5.587
59.4	0.7412	6.171	64.4	0.7223	6.013	69.4	0.7043	5.863	74.4	0.6872	5.720	79.4	0.6709	5.584
59.5	0.7408	6.167	64.5	0.7219	6.010	69.5	0.7040	5.860	74.5	0.6869	5.717	79.5	0.6706	5.582
59.6	0.7405	6.164	64.6	0.7216	6.007	69.6	0.7036	5.857	74.6	0.6866	5.715	79.6	0.6703	5.579
59.7	0.7401	6.161	64.7	0.7212	6.004	69.7	0.7033	5.854	74.7	0.6862	5.712	79.7	0.6700	5.576
59.8	0.7397	6.158	64.8	0.7208	6.000	69.8	0.7029	5.851	74.8	0.6859	5.709	79.8	0.6697	5.574
59.9	0.7393	6.154	64.9	0.7205	5.997	69.9	0.7026	5.848	74.9	0.6856	5.706	79.9	0.6693	5.571
												80.0	0.6690	5.569

°API GRAVITY, SPECIFIC GRAVITY (60°/60°) AND POUNDS PER GALLON (Continued)

30 TO 55° API

°API	SPECIFIC GRAVITY	POUNDS PER GALLON	°API	SPECIFIC GRAVITY	POUNDS PER GALLON	°API	SPECIFIC GRAVITY	POUNDS PER GALLON	°API	SPECIFIC GRAVITY	POUNDS PER GALLON	°API	SPECIFIC GRAVITY	POUNDS PER GALLON
30.0	0.8762	7.296	35.0	0.8499	7.076	40.0	0.8251	6.870	45.0	0.8017	6.675	50.0	0.7796	6.491
30.1	0.8756	7.291	35.1	0.8493	7.072	40.1	0.8246	6.866	45.1	0.8012	6.671	50.1	0.7792	6.487
30.2	0.8751	7.287	35.2	0.8488	7.068	40.2	0.8241	6.862	45.2	0.8008	6.667	50.2	0.7788	6.483
30.3	0.8745	7.282	35.3	0.8483	7.064	40.3	0.8236	6.858	45.3	0.8003	6.663	50.3	0.7783	6.480
30.4	0.8740	7.278	35.4	0.8478	7.059	40.4	0.8232	6.854	45.4	0.7999	6.660	50.4	0.7779	6.476
30.5	0.8735	7.273	35.5	0.8473	7.055	40.5	0.8227	6.850	45.5	0.7994	6.656	50.5	0.7775	6.473
30.6	0.8729	7.269	35.6	0.8468	7.051	40.6	0.8222	6.846	45.6	0.7990	6.652	50.6	0.7770	6.469
30.7	0.8724	7.264	35.7	0.8463	7.047	40.7	0.8217	6.842	45.7	0.7985	6.648	50.7	0.7766	6.466
30.8	0.8718	7.260	35.8	0.8458	7.042	40.8	0.8212	6.838	45.8	0.7981	6.645	50.8	0.7762	6.462
30.9	0.8713	7.255	35.9	0.8453	7.038	40.9	0.8208	6.834	45.9	0.7976	6.641	50.9	0.7758	6.459
31.0	0.8708	7.251	36.0	0.8448	7.034	41.0	0.8203	6.830	46.0	0.7972	6.637	51.0	0.7753	6.455
31.1	0.8702	7.246	36.1	0.8443	7.030	41.1	0.8198	6.826	46.1	0.7967	6.633	51.1	0.7749	6.451
31.2	0.8697	7.242	36.2	0.8438	7.026	41.2	0.8193	6.822	46.2	0.7963	6.630	51.2	0.7745	6.448
31.3	0.8692	7.237	36.3	0.8433	7.021	41.3	0.8189	6.818	46.3	0.7958	6.626	51.3	0.7741	6.444
31.4	0.8686	7.233	36.4	0.8428	7.017	41.4	0.8184	6.814	46.4	0.7954	6.622	51.4	0.7736	6.441
31.5	0.8681	7.228	36.5	0.8423	7.013	41.5	0.8179	6.810	46.5	0.7949	6.618	51.5	0.7732	6.437
31.6	0.8676	7.224	36.6	0.8418	7.009	41.6	0.8174	6.806	46.6	0.7945	6.615	51.6	0.7728	6.434
31.7	0.8670	7.220	36.7	0.8413	7.005	41.7	0.8170	6.802	46.7	0.7941	6.611	51.7	0.7724	6.430
31.8	0.8665	7.215	36.8	0.8408	7.000	41.8	0.8165	6.798	46.8	0.7936	6.607	51.8	0.7720	6.427
31.9	0.8660	7.211	36.9	0.8403	6.996	41.9	0.8160	6.794	46.9	0.7932	6.604	51.9	0.7715	6.423
32.0	0.8654	7.206	37.0	0.8398	6.992	42.0	0.8156	6.790	47.0	0.7927	6.600	52.0	0.7711	6.420
32.1	0.8649	7.202	37.1	0.8393	6.988	42.1	0.8151	6.786	47.1	0.7923	6.596	52.1	0.7707	6.416
32.2	0.8644	7.197	37.2	0.8388	6.984	42.2	0.8146	6.783	47.2	0.7918	6.592	52.2	0.7703	6.413
32.3	0.8639	7.193	37.3	0.8383	6.980	42.3	0.8142	6.779	47.3	0.7914	6.589	52.3	0.7699	6.409
32.4	0.8633	7.189	37.4	0.8378	6.976	42.4	0.8137	6.775	47.4	0.7909	6.585	52.4	0.7694	6.406
32.5	0.8628	7.184	37.5	0.8373	6.971	42.5	0.8132	6.771	47.5	0.7905	6.581	52.5	0.7690	6.402
32.6	0.8623	7.180	37.6	0.8368	6.967	42.6	0.8128	6.767	47.6	0.7901	6.578	52.6	0.7686	6.399
32.7	0.8618	7.176	37.7	0.8363	6.963	42.7	0.8123	6.763	47.7	0.7896	6.574	52.7	0.7682	6.395
32.8	0.8612	7.171	37.8	0.8358	6.959	42.8	0.8118	6.759	47.8	0.7892	6.570	52.8	0.7678	6.392
32.9	0.8607	7.167	37.9	0.8353	6.955	42.9	0.8114	6.755	47.9	0.7887	6.567	52.9	0.7674	6.388
33.0	0.8602	7.162	38.0	0.8348	6.951	43.0	0.8109	6.751	48.0	0.7883	6.563	53.0	0.7669	6.385
33.1	0.8597	7.158	38.1	0.8343	6.947	43.1	0.8104	6.748	48.1	0.7879	6.559	53.1	0.7665	6.381
33.2	0.8591	7.154	38.2	0.8338	6.943	43.2	0.8100	6.744	48.2	0.7874	6.556	53.2	0.7661	6.378
33.3	0.8586	7.149	38.3	0.8333	6.939	43.3	0.8095	6.740	48.3	0.7870	6.552	53.3	0.7657	6.375
33.4	0.8581	7.145	38.4	0.8328	6.934	43.4	0.8090	6.736	48.4	0.7865	6.548	53.4	0.7563	6.371
33.5	0.8576	7.141	38.5	0.8324	6.930	43.5	0.8086	6.732	48.5	0.7861	6.545	53.5	0.7649	6.368
33.6	0.8571	7.136	38.6	0.8319	6.926	43.6	0.8081	6.728	48.6	0.7857	6.541	53.6	0.7645	6.364
33.7	0.8565	7.132	38.7	0.8314	6.922	43.7	0.8076	6.724	48.7	0.7852	6.538	53.7	0.7640	6.361
33.8	0.8560	7.128	38.8	0.8309	6.918	43.8	0.8072	6.721	48.8	0.7848	6.534	53.8	0.7636	6.357
33.9	0.8555	7.123	38.9	0.8304	6.914	43.9	0.8067	6.717	48.9	0.7844	6.530	53.9	0.7632	6.354
34.0	0.8550	7.119	39.0	0.8299	6.910	44.0	0.8063	6.713	49.0	0.7839	6.527	54.0	0.7628	6.350
34.1	0.8545	7.115	39.1	0.8294	6.906	44.1	0.8058	6.709	49.1	0.7835	6.523	54.1	0.7624	6.347
34.2	0.8540	7.110	39.2	0.8289	6.902	44.2	0.8054	6.705	49.2	0.7831	6.519	54.2	0.7620	6.344
34.3	0.8534	7.106	39.3	0.8285	6.898	44.3	0.8049	6.701	49.3	0.7826	6.516	54.3	0.7616	6.340
34.4	0.8529	7.102	39.4	0.8280	6.894	44.4	0.8044	6.698	49.4	0.7822	6.512	54.4	0.7612	6.337
34.5	0.8524	7.098	39.5	0.8275	6.890	44.5	0.8040	6.694	49.5	0.7818	6.509	54.5	0.7608	6.333
34.6	0.8519	7.093	39.6	0.8270	6.886	44.6	0.8035	6.690	49.6	0.7813	6.505	54.6	0.7603	6.330
34.7	0.8514	7.089	39.7	0.8265	6.882	44.7	0.8031	6.686	49.7	0.7809	6.501	54.7	0.7599	6.327
34.8	0.8509	7.085	39.8	0.8260	6.878	44.8	0.8026	6.682	49.8	0.7805	6.498	54.8	0.7595	6.323
34.9	0.8504	7.081	39.9	0.8256	6.874	44.9	0.8022	6.679	49.9	0.7800	6.494	54.9	0.7591	6.320
												55.0	0.7587	6.316

°API GRAVITY, SPECIFIC GRAVITY (60°/60°) AND POUNDS PER GALLON (Continued)

55 TO 80°API

°API	SPECIFIC GRAVITY	POUNDS PER GALLON	°API	SPECIFIC GRAVITY	POUNDS PER GALLON	°API	SPECIFIC GRAVITY	POUNDS PER GALLON	°API	SPECIFIC GRAVITY	POUNDS PER GALLON	°API	SPECIFIC GRAVITY	POUNDS PER GALLON
55.0	0.7587	6.316	60.0	0.7389	6.151	65.0	0.7201	5.994	70.0	0.7022	5.845	75.0	0.6852	5.704
55.1	0.7583	6.313	60.1	0.7385	6.148	65.1	0.7197	5.991	70.1	0.7019	5.842	75.1	0.6849	5.701
55.2	0.7579	6.310	60.2	0.7381	6.145	65.2	0.7194	5.988	70.2	0.7015	5.840	75.2	0.6846	5.698
55.3	0.7575	6.306	60.3	0.7377	6.141	65.3	0.7190	5.985	70.3	0.7012	5.837	75.3	0.6842	5.695
55.4	0.7571	6.303	60.4	0.7374	6.138	65.4	0.7186	5.982	70.4	0.7008	5.834	75.4	0.6839	5.693
55.5	0.7567	6.299	60.5	0.7370	6.135	65.5	0.7183	5.979	70.5	0.7005	5.831	75.5	0.6836	5.690
55.6	0.7563	6.296	60.6	0.7366	6.132	65.6	0.7179	5.976	70.6	0.7001	5.828	75.6	0.6832	5.687
55.7	0.7559	6.293	60.7	0.7362	6.129	65.7	0.7175	5.973	70.7	0.6998	5.825	75.7	0.6829	5.684
55.8	0.7555	6.289	60.8	0.7358	6.126	65.8	0.7172	5.970	70.8	0.6995	5.822	75.8	0.6826	5.682
55.9	0.7551	6.286	60.9	0.7354	6.122	65.9	0.7168	5.967	70.9	0.6991	5.819	75.9	0.6823	5.679
56.0	0.7547	6.283	61.0	0.7351	6.119	66.0	0.7165	5.964	71.0	0.6988	5.816	76.0	0.6819	5.676
56.1	0.7543	6.279	61.1	0.7347	6.116	66.1	0.7161	5.961	71.1	0.6984	5.814	76.1	0.6816	5.673
56.2	0.7539	6.276	61.2	0.7343	6.113	66.2	0.7157	5.958	71.2	0.6981	5.811	76.2	0.6813	5.671
56.3	0.7535	6.273	61.3	0.7339	6.110	66.3	0.7154	5.955	71.3	0.6977	5.808	76.3	0.6809	5.668
56.4	0.7531	6.269	61.4	0.7335	6.106	66.4	0.7150	5.952	71.4	0.6974	5.805	76.4	0.6806	5.665
56.5	0.7527	6.266	61.5	0.7332	6.103	66.5	0.7146	5.949	71.5	0.6970	5.802	76.5	0.6803	5.662
56.6	0.7523	6.263	61.6	0.7328	6.100	66.6	0.7143	5.946	71.6	0.6967	5.799	76.6	0.6800	5.660
56.7	0.7519	6.259	61.7	0.7324	6.097	66.7	0.7139	5.943	71.7	0.6964	5.796	76.7	0.6796	5.657
56.8	0.7515	6.256	61.8	0.7320	6.094	66.8	0.7136	5.940	71.8	0.6960	5.794	76.8	0.6793	5.654
56.9	0.7511	6.253	61.9	0.7316	6.091	66.9	0.7132	5.937	71.9	0.6957	5.791	76.9	0.6790	5.651
57.0	0.7507	6.249	62.0	0.7313	6.087	67.0	0.7128	5.934	72.0	0.6953	5.788	77.0	0.6787	5.649
57.1	0.7503	6.246	62.1	0.7309	6.084	67.1	0.7125	5.931	72.1	0.6950	5.785	77.1	0.6783	5.646
57.2	0.7499	6.243	62.2	0.7305	6.081	67.2	0.7121	5.928	72.2	0.6946	5.782	77.2	0.6780	5.643
57.3	0.7495	6.239	62.3	0.7301	6.078	67.3	0.7118	5.925	72.3	0.6943	5.779	77.3	0.6777	5.641
57.4	0.7491	6.236	62.4	0.7298	6.075	67.4	0.7114	5.922	72.4	0.6940	5.776	77.4	0.6774	5.638
57.5	0.7487	6.233	62.5	0.7294	6.072	67.5	0.7111	5.919	72.5	0.6936	5.774	77.5	0.6770	5.635
57.6	0.7483	6.229	62.6	0.7290	6.069	67.6	0.7107	5.916	72.6	0.6933	5.771	77.6	0.6767	5.633
57.7	0.7479	6.226	62.7	0.7286	6.065	67.7	0.7103	5.913	72.7	0.6929	5.768	77.7	0.6764	5.630
57.8	0.7475	6.223	62.8	0.7283	6.062	67.8	0.7100	5.910	72.8	0.6926	5.765	77.8	0.6761	5.627
57.9	0.7471	6.219	62.9	0.7279	6.059	67.9	0.7096	5.907	72.9	0.6923	5.762	77.9	0.6757	5.624
58.0	0.7467	6.216	63.0	0.7275	6.056	68.0	0.7093	5.904	73.0	0.6919	5.759	78.0	0.6754	5.622
58.1	0.7463	6.213	63.1	0.7271	6.053	68.1	0.7089	5.901	73.1	0.6916	5.757	78.1	0.6751	5.619
58.2	0.7459	6.210	63.2	0.7268	6.050	68.2	0.7086	5.898	73.2	0.6913	5.754	78.2	0.6748	5.616
58.3	0.7455	6.206	63.3	0.7264	6.047	68.3	0.7082	5.895	73.3	0.6909	5.751	78.3	0.6745	5.614
58.4	0.7451	6.203	63.4	0.7260	6.044	68.4	0.7079	5.892	73.4	0.6906	5.748	78.4	0.6741	5.611
58.5	0.7447	6.200	63.5	0.7256	6.041	68.5	0.7075	5.889	73.5	0.6902	5.745	78.5	0.6738	5.608
58.6	0.7443	6.197	63.6	0.7253	6.037	68.6	0.7071	5.886	73.6	0.6899	5.743	78.6	0.6735	5.606
58.7	0.7440	6.193	63.7	0.7249	6.034	68.7	0.7068	5.883	73.7	0.6896	5.740	78.7	0.6732	5.603
58.8	0.7436	6.190	63.8	0.7245	6.031	68.8	0.7064	5.880	73.8	0.6892	5.737	78.8	0.6728	5.600
58.9	0.7432	6.187	63.9	0.7242	6.028	68.9	0.7061	5.878	73.9	0.6889	5.734	78.9	0.6725	5.598
59.0	0.7428	6.183	64.0	0.7238	6.025	69.0	0.7057	5.875	74.0	0.6886	5.731	79.0	0.6722	5.595
59.1	0.7424	6.180	64.1	0.7234	6.022	69.1	0.7054	5.872	74.1	0.6882	5.729	79.1	0.6719	5.592
59.2	0.7420	6.177	64.2	0.7230	6.019	69.2	0.7050	5.869	74.2	0.6879	5.726	79.2	0.6716	5.590
59.3	0.7416	6.174	64.3	0.7227	6.016	69.3	0.7047	5.866	74.3	0.6876	5.723	79.3	0.6713	5.587
59.4	0.7412	6.171	64.4	0.7223	6.013	69.4	0.7043	5.863	74.4	0.6872	5.720	79.4	0.6709	5.584
59.5	0.7408	6.167	64.5	0.7219	6.010	69.5	0.7040	5.860	74.5	0.6869	5.717	79.5	0.6706	5.582
59.6	0.7405	6.164	64.6	0.7216	6.007	69.6	0.7036	5.857	74.6	0.6866	5.715	79.6	0.6703	5.579
59.7	0.7401	6.161	64.7	0.7212	6.004	69.7	0.7033	5.854	74.7	0.6862	5.712	79.7	0.6700	5.576
59.8	0.7397	6.158	64.8	0.7208	6.000	69.8	0.7029	5.851	74.8	0.6859	5.709	79.8	0.6697	5.574
59.9	0.7393	6.154	64.9	0.7205	5.997	69.9	0.7026	5.848	74.9	0.6856	5.706	79.9	0.6693	5.571
												80.0	0.6690	5.569

°API GRAVITY, SPECIFIC GRAVITY (60°/60°) AND POUNDS PER GALLON (Continued)

80 TO 100°API

°API	SPECIFIC GRAVITY	POUNDS PER GALLON	°API	SPECIFIC GRAVITY	POUNDS PER GALLON	°API	SPECIFIC GRAVITY	POUNDS PER GALLON	°API	SPECIFIC GRAVITY	POUNDS PER GALLON
80.0	0.6690	5.569	85.0	0.6536	5.440	90.0	0.6388	5.317	95.0	0.6247	5.199
80.1	0.6687	5.566	85.1	0.6533	5.437	90.1	0.6385	5.314	95.1	0.6244	5.197
80.2	0.6684	5.563	85.2	0.6530	5.435	90.2	0.6383	5.312	95.2	0.6242	5.194
80.3	0.6681	5.561	85.3	0.6527	5.432	90.3	0.6380	5.309	95.3	0.6239	5.192
80.4	0.6678	5.558	85.4	0.6524	5.430	90.4	0.6377	5.307	95.4	0.6236	5.190
80.5	0.6675	5.555	85.5	0.6521	5.427	90.5	0.6374	5.305	95.5	0.6233	5.188
80.6	0.6671	5.553	85.6	0.6518	5.425	90.6	0.6371	5.302	95.6	0.6231	5.185
80.7	0.6668	5.550	85.7	0.6515	5.422	90.7	0.6368	5.300	95.7	0.6228	5.183
80.8	0.6665	5.547	85.8	0.6512	5.420	90.8	0.6365	5.297	95.8	0.6225	5.181
80.9	0.6662	5.545	85.9	0.6509	5.417	90.9	0.6362	5.295	95.9	0.6223	5.178
81.0	0.6659	5.542	86.0	0.6506	5.415	91.0	0.6360	5.293	96.0	0.6220	5.176
81.1	0.6656	5.540	86.1	0.6503	5.412	91.1	0.6357	5.290	96.1	0.6217	5.174
81.2	0.6653	5.537	86.2	0.6500	5.410	91.2	0.6354	5.288	96.2	0.6214	5.172
81.3	0.6649	5.534	86.3	0.6497	5.407	91.3	0.6351	5.286	96.3	0.6212	5.169
81.4	0.6646	5.532	86.4	0.6494	5.405	91.4	0.6348	5.283	96.4	0.6209	5.167
81.5	0.6643	5.529	86.5	0.6491	5.402	91.5	0.6345	5.281	96.5	0.6206	5.165
81.6	0.6640	5.527	86.6	0.6488	5.400	91.6	0.6342	5.278	96.6	0.6203	5.163
81.7	0.6637	5.524	86.7	0.6485	5.397	91.7	0.6340	5.276	96.8	0.6201	5.160
81.8	0.6634	5.521	86.8	0.6482	5.395	91.8	0.6337	5.274	96.8	0.6198	5.158
81.9	0.6631	5.519	86.9	0.6479	5.392	91.9	0.6334	5.271	96.9	0.6195	5.156
82.0	0.6628	5.516	87.0	0.6476	5.390	92.0	0.6331	5.269	97.0	0.6193	5.153
82.1	0.6625	5.514	87.1	0.6473	5.387	92.1	0.6328	5.267	97.1	0.6190	5.151
82.2	0.6621	5.511	87.2	0.6470	5.385	92.2	0.6325	5.264	97.2	0.6187	5.149
82.3	0.6618	5.508	87.3	0.6467	5.382	92.3	0.6323	5.262	97.3	0.6184	5.147
82.4	0.6615	5.506	87.4	0.6464	5.380	92.4	0.6320	5.260	97.4	0.6182	5.144
82.5	0.6612	5.503	87.5	0.6461	5.377	92.5	0.6317	5.257	97.5	0.6179	5.142
82.6	0.6609	5.501	87.6	0.6458	5.375	92.6	0.6314	5.255	97.6	0.6176	5.140
82.7	0.6606	5.498	87.7	0.6455	5.373	92.7	0.6311	5.252	97.7	0.6174	5.138
82.8	0.6603	5.496	87.8	0.6452	5.370	92.8	0.6309	5.250	97.8	0.6171	5.135
82.9	0.6600	5.493	87.9	0.6449	5.368	92.9	0.6306	5.248	97.9	0.6168	5.133
83.0	0.6597	5.490	88.0	0.6446	5.365	93.0	0.6303	5.245	98.0	0.6166	5.131
83.1	0.6594	5.488	88.1	0.6444	5.363	93.1	0.6300	5.243	98.1	0.6163	5.129
83.2	0.6591	5.485	88.2	0.6441	5.360	93.2	0.6297	5.241	98.2	0.6160	5.126
83.3	0.6588	5.483	88.3	0.6438	5.358	93.3	0.6294	5.238	98.3	0.6158	5.124
83.4	0.6584	5.480	88.4	0.6435	5.355	93.4	0.6292	5.236	98.4	0.6155	5.122
83.5	0.6581	5.478	88.5	0.6432	5.353	93.5	0.6289	5.234	98.5	0.6152	5.120
83.6	0.6578	5.475	88.6	0.6429	5.351	93.6	0.6286	5.231	98.6	0.6150	5.118
83.7	0.6575	5.473	88.7	0.6426	5.348	93.7	0.6283	5.229	98.7	0.6147	5.115
83.8	0.6572	5.470	88.8	0.6423	5.346	93.8	0.6281	5.227	98.8	0.6144	5.113
83.9	0.6569	5.468	88.9	0.6420	5.343	93.9	0.6278	5.224	98.9	0.6141	5.111
84.0	0.6566	5.465	89.0	0.6417	5.341	94.0	0.6275	5.222	99.0	0.6139	5.109
84.1	0.6563	5.462	89.1	0.6414	5.338	94.1	0.6272	5.220	99.1	0.6136	5.106
84.2	0.6560	5.460	89.2	0.6411	5.336	94.2	0.6269	5.218	99.2	0.6134	5.104
84.3	0.6557	5.457	89.3	0.6409	5.334	94.3	0.6267	5.215	99.3	0.6131	5.102
84.4	0.6554	5.455	89.4	0.6406	5.331	94.4	0.6264	5.213	99.4	0.6128	5.100
84.5	0.6551	5.452	89.5	0.6403	5.329	94.5	0.6261	5.211	99.5	0.6126	5.098
84.6	0.6548	5.450	89.6	0.6400	5.326	94.6	0.6258	5.208	99.6	0.6123	5.095
84.7	0.6545	5.447	89.7	0.6397	5.324	94.7	0.6256	5.206	99.7	0.6120	5.093
84.8	0.6542	5.445	89.8	0.6394	5.321	94.8	0.6253	5.204	99.8	0.6118	5.091
84.9	0.6539	5.442	89.9	0.6391	5.319	94.9	0.6250	5.201	99.9	0.6115	5.089
									100.0	0.6112	5.087

°API GRAVITY, SPECIFIC GRAVITY (60°/60°)
AND POUNDS PER GALLON (Continued)

100 TO 200°API

°API	SPECIFIC GRAVITY	POUNDS PER GALLON	°API	SPECIFIC GRAVITY	POUNDS PER GALLON	°API	SPECIFIC GRAVITY	POUNDS PER GALLON	°API	SPECIFIC GRAVITY	POUNDS PER GALLON	°API	SPECIFIC GRAVITY	POUNDS PER GALLON	°API	SPECIFIC GRAVITY	POUNDS PER GALLON
100.0	0.6112	5.087	120.0	0.5626	4.681	140.0	0.5212	4.336	160.0	0.4854	4.038	180.0	0.4543	3.778
100.5	0.6099	5.076	120.5	0.5615	4.672	140.5	0.5202	4.328	160.5	0.4846	4.031	180.5	0.4535	3.772
101.0	0.6086	5.065	121.0	0.5604	4.663	141.0	0.5193	4.320	161.0	0.4838	4.024	181.0	0.4528	3.766
101.5	0.6073	5.054	121.5	0.5593	4.654	141.5	0.5183	4.312	161.5	0.4829	4.017	181.5	0.4521	3.760
102.0	0.6060	5.043	122.0	0.5582	4.644	142.0	0.5174	4.304	162.0	0.4821	4.010	182.0	0.4514	3.754
102.5	0.6047	5.032	122.5	0.5571	4.635	142.5	0.5164	4.296	162.5	0.4813	4.003	182.5	0.4506	3.748
103.0	0.6034	5.022	123.0	0.5560	4.626	143.0	0.5155	4.288	163.0	0.4805	3.996	183.0	0.4499	3.742
103.5	0.6021	5.011	123.5	0.5549	4.617	143.5	0.5145	4.281	163.5	0.4797	3.990	183.5	0.4492	3.736
104.0	0.6009	5.000	124.0	0.5538	4.608	144.0	0.5136	4.273	164.0	0.4789	3.983	184.0	0.4485	3.730
104.5	0.5996	4.990	124.5	0.5527	4.599	144.5	0.5127	4.265	164.5	0.4780	3.976	184.5	0.4478	3.724
105.0	0.5983	4.979	125.0	0.5517	4.590	145.0	0.5118	4.257	165.0	0.4772	3.969	185.0	0.4471	3.718
105.5	0.5970	4.968	125.5	0.5506	4.581	145.5	0.5108	4.250	165.5	0.4764	3.963	185.5	0.4464	3.712
106.0	0.5958	4.958	126.0	0.5495	4.572	146.0	0.5099	4.242	166.0	0.4756	3.956	186.0	0.4457	3.706
106.5	0.5945	4.948	126.5	0.5485	4.563	146.5	0.5090	4.234	166.5	0.4748	3.949	186.5	0.4450	3.700
107.0	0.5933	4.937	127.0	0.5474	4.554	147.0	0.5081	4.227	167.0	0.4740	3.943	187.0	0.4443	3.695
107.5	0.5921	4.927	127.5	0.5463	4.546	147.5	0.5072	4.219	167.5	0.4732	3.936	187.5	0.4436	3.689
108.0	0.5908	4.916	128.0	0.5453	4.537	148.0	0.5063	4.211	168.0	0.4725	3.930	188.0	0.4429	3.683
108.5	0.5896	4.906	128.5	0.5442	4.528	148.5	0.5054	4.204	168.5	0.4717	3.923	188.5	0.4422	3.677
109.0	0.5884	4.896	129.0	0.5432	4.519	149.0	0.5045	4.196	169.0	0.4709	3.916	189.0	0.4415	3.671
109.5	0.5871	4.886	129.5	0.5421	4.511	149.5	0.5036	4.189	169.5	0.4701	3.910	189.5	0.4408	3.666
110.0	0.5859	4.876	130.0	0.5411	4.502	150.0	0.5027	4.181	170.0	0.4693	3.903	190.0	0.4401	3.660
110.5	0.5847	4.866	130.5	0.5401	4.493	150.5	0.5018	4.174	170.5	0.4685	3.897	190.5	0.4394	3.654
111.0	0.5835	4.856	131.0	0.5390	4.485	151.0	0.5009	4.167	171.0	0.4678	3.891	191.0	0.4388	3.649
111.5	0.5823	4.846	131.5	0.5380	4.476	151.5	0.5000	4.159	171.5	0.4670	3.884	191.5	0.4381	3.643
112.0	0.5811	4.836	132.0	0.5370	4.468	152.0	0.4991	4.152	172.0	0.4662	3.878	192.0	0.4374	3.637
112.5	0.5799	4.826	132.5	0.5360	4.459	152.5	0.4982	4.145	172.5	0.4655	3.871	192.5	0.4367	3.632
113.0	0.5787	4.816	133.0	0.5350	4.451	153.0	0.4974	4.137	173.0	0.4647	3.865	193.0	0.4361	3.626
113.5	0.5776	4.806	133.5	0.5340	4.442	153.5	0.4965	4.130	173.5	0.4639	3.859	193.5	0.4354	3.621
114.0	0.5764	4.796	134.0	0.5330	4.434	154.0	0.4956	4.123	174.0	0.4632	3.852	194.0	0.4347	3.615
114.5	0.5752	4.786	134.5	0.5320	4.426	154.5	0.4948	4.116	174.5	0.4624	3.846	194.5	0.4341	3.609
115.0	0.5740	4.777	135.0	0.5310	4.417	155.0	0.4939	4.108	175.0	0.4617	3.840	195.0	0.4334	3.604
115.5	0.5729	4.767	135.5	0.5300	4.409	155.5	0.4930	4.101	175.5	0.4609	3.833	195.5	0.4327	3.598
116.0	0.5717	4.757	136.0	0.5290	4.401	156.0	0.4922	4.094	176.0	0.4602	3.827	196.0	0.4321	3.593
116.5	0.5706	4.748	136.5	0.5280	4.393	156.5	0.4913	4.087	176.5	0.4594	3.821	196.5	0.4314	3.587
117.0	0.5694	4.738	137.0	0.5270	4.384	157.0	0.4905	4.080	177.0	0.4587	3.815	197.0	0.4307	3.582
117.5	0.5683	4.729	137.5	0.5260	4.376	157.5	0.4896	4.073	177.5	0.4579	3.808	197.5	0.4301	3.576
118.0	0.5671	4.719	138.0	0.5250	4.368	158.0	0.4888	4.066	178.0	0.4572	3.802	198.0	0.4294	3.571
118.5	0.5660	4.710	138.5	0.5241	4.360	158.5	0.4879	4.059	178.5	0.4565	3.796	198.5	0.4288	3.566
119.0	0.5649	4.700	139.0	0.5231	4.352	159.0	0.4871	4.052	179.0	0.4557	3.790	199.0	0.4281	3.560
119.5	0.5637	4.691	139.5	0.5221	4.344	159.5	0.4863	4.045	179.5	0.4550	3.784	199.5	0.4275	3.555
												200.0	0.4269	3.549

°API GRAVITY, SPECIFIC GRAVITY (60°/60°)
AND POUNDS PER GALLON (Continued)

200 TO 300°API

°API	SPECIFIC GRAVITY	POUNDS PER GALLON	°API	SPECIFIC GRAVITY	POUNDS PER GALLON	°API	SPECIFIC GRAVITY	POUNDS PER GALLON	°API	SPECIFIC GRAVITY	POUNDS PER GALLON	°API	SPECIFIC GRAVITY	POUNDS PER GALLON
200.0	0.4269	3.549	220.0	0.4026	3.347	240.0	0.3809	3.166	260.0	0.3614	3.004	280.0	0.3439	2.857
200.5	0.4262	3.544	220.5	0.4020	3.342	240.5	0.3804	3.162	260.5	0.3610	3.000	280.5	0.3434	2.854
201.0	0.4256	3.539	221.0	0.4014	3.337	241.0	0.3799	3.158	261.0	0.3605	2.996	281.0	0.3430	2.850
201.5	0.4249	3.533	221.5	0.4009	3.333	241.5	0.3794	3.153	261.5	0.3601	2.992	281.5	0.3426	2.847
202.0	0.4243	3.528	222.0	0.4003	3.328	242.0	0.3789	3.149	262.0	0.3596	2.989	282.0	0.3422	2.844
202.5	0.4237	3.523	222.5	0.3997	3.323	242.5	0.3783	3.145	262.5	0.3591	2.985	282.5	0.3418	2.840
203.0	0.4230	3.517	223.0	0.3992	3.318	243.0	0.3778	3.141	263.0	0.3587	2.981	283.0	0.3414	2.837
203.5	0.4224	3.512	223.5	0.3986	3.314	243.5	0.3773	3.136	263.5	0.3582	2.977	283.5	0.3410	2.833
204.0	0.4218	3.507	224.0	0.3980	3.308	244.0	0.3768	3.132	264.0	0.3578	2.873	284.0	0.3406	2.830
204.5	0.4211	3.502	224.5	0.3975	3.304	244.5	0.3763	3.128	264.5	0.3573	2.970	284.5	0.3401	2.826
205.0	0.4205	3.496	225.0	0.3969	3.300	245.0	0.3758	3.124	265.0	0.3569	2.966	285.0	0.3397	2.823
205.5	0.4199	3.491	225.5	0.3964	3.295	245.5	0.3753	3.120	265.5	0.3564	2.962	285.5	0.3393	2.820
206.0	0.4193	3.486	226.0	0.3958	3.290	246.0	0.3748	3.116	266.0	0.3560	2.958	286.0	0.3389	2.816
206.5	0.4186	3.481	226.5	0.3953	3.286	246.5	0.3743	3.112	266.5	0.3555	2.955	286.5	0.3385	2.813
207.0	0.4180	3.476	227.0	0.3947	3.281	247.0	0.3738	3.107	267.0	0.3551	2.951	287.0	0.3381	2.809
207.5	0.4174	3.471	227.5	0.3942	3.277	247.5	0.3734	3.103	267.5	0.3546	2.947	287.5	0.3377	2.806
208.0	0.4168	3.465	228.0	0.3936	3.272	248.0	0.3729	3.099	268.0	0.3542	2.944	288.0	0.3373	2.803
208.5	0.4162	3.460	228.5	0.3931	3.268	248.5	0.3724	3.095	268.5	0.3538	2.940	288.5	0.3369	2.799
209.0	0.4156	3.455	229.0	0.3925	3.263	249.0	0.3719	3.091	269.0	0.3533	2.936	289.0	0.3365	2.796
209.5	0.4150	3.450	229.5	0.3920	3.259	249.5	0.3714	3.087	269.5	0.3529	2.932	289.5	0.3361	2.793
210.0	0.4144	3.445	230.0	0.3914	3.254	250.0	0.3709	3.083	270.0	0.3524	2.929	290.0	0.3357	2.789
210.5	0.4137	3.440	230.5	0.3909	3.249	250.5	0.3704	3.079	270.5	0.3520	2.925	290.5	0.3353	2.786
211.0	0.4131	3.435	231.0	0.3903	3.245	251.0	0.3699	3.075	271.0	0.3516	2.922	291.0	0.3349	2.783
211.5	0.4125	3.430	231.5	0.3898	3.240	251.5	0.3695	3.071	271.5	0.3511	2.918	291.5	0.3345	2.779
212.0	0.4119	3.425	232.0	0.3893	3.236	252.0	0.3690	3.067	272.0	0.3507	2.914	292.0	0.3341	2.776
212.5	0.4113	3.420	232.5	0.3887	3.232	252.5	0.3685	3.063	272.5	0.3503	2.911	292.5	0.3337	2.773
213.0	0.4107	3.415	233.0	0.3882	3.227	253.0	0.3680	3.059	273.0	0.3498	2.907	293.0	0.3333	2.770
213.5	0.4101	3.410	233.5	0.3877	3.223	253.5	0.3675	3.055	273.5	0.3494	2.903	293.5	0.3329	2.766
214.0	0.4096	3.405	234.0	0.3871	3.218	254.0	0.3671	3.051	274.0	0.3490	2.900	294.0	0.3326	2.763
214.5	0.4090	3.400	234.5	0.3866	3.214	254.5	0.3666	3.047	274.5	0.3485	2.896	294.5	0.3322	2.760
215.0	0.4084	3.395	235.0	0.3861	3.209	255.0	0.3661	3.043	275.0	0.3481	2.893	295.0	0.3318	2.757
215.5	0.4078	3.390	235.5	0.3856	3.205	255.5	0.3656	3.039	275.5	0.3477	2.889	295.5	0.3314	2.753
216.0	0.4072	3.385	236.0	0.3850	3.201	256.0	0.3652	3.035	276.0	0.3472	2.886	296.0	0.3310	2.750
216.5	0.4066	3.381	236.5	0.3845	3.196	256.5	0.3647	3.031	276.5	0.3468	2.882	296.5	0.3306	2.747
217.0	0.4060	3.376	237.0	0.3840	3.192	257.0	0.3642	3.027	277.0	0.3464	2.878	297.0	0.3302	2.744
217.5	0.4054	3.371	237.5	0.3835	3.188	257.5	0.3638	3.023	277.5	0.3460	2.875	297.5	0.3298	2.740
218.0	0.4049	3.366	238.0	0.3830	3.183	258.0	0.3633	3.019	278.0	0.3455	2.871	298.0	0.3295	2.737
218.5	0.4043	3.361	238.5	0.3824	3.179	258.5	0.3628	3.015	278.5	0.3451	2.868	298.5	0.3291	2.734
219.0	0.4037	3.356	239.0	0)3819	3.175	259.0	0.3624	3.012	279.0	0.3447	2.864	299.0	0.3287	2.731
219.5	0.4031	3.352	239.5	0.3814	3.170	259.5	0.3619	3.008	279.5	0.3443	2.861	299.5	0.3283	2.727
												300.0	0.3279	2.724

Index